72 Beautiful Galaxies

© 2016 by Stephen Perrenod

ISBN: 978-0-692-77020-7

Photo "72 Thonglor"

by Stephen Perrenod © 2016

Cover image: Andromeda Galaxy (M31) taken in ultraviolet light

by the GALEX satellite.

"Andromeda galaxy" by NASA/JPL/California Institute of Technology

This image or video was catalogued by Jet Propulsion Laboratory of the United States National Aeronautics and Space Administration (NASA) under Photo ID: PIA04921. Public domain.

Dedication

This book is dedicated to my wife of 28 years, Yuriko Fukazawa, who loved astronomy as well.

Foreword

Somewhere on this planet long ago, our distant ancestors first looked up to the heavens and began to wonder. Was there something more than themselves, their tribes, and their territories?

Without the means to better see the stars, we were at first captives of wonder. Constellations were drawn as metaphors for things we could understand.

And as we began to study celestial movements, we looked to the constellations for omens of the future. How ironic, then, that the ancient light from those stars had taken millions, even billions, of years to reach our sight.

Galaxies first appeared to us as single points of light or faint, fuzzy patches. But looking deeper with modern telescopes, we can see them for the tribes that they really are. As we learn in this book, these galaxies form the wheels of creation -- not only of new stars and planets, but the core elements that make up our very being.

Thanks to the works of Dr. Stephen Perrenod and other astrophysicists, we are no longer captives of wonder. The stories told around campfires have given way to descriptions of a different type of tribe altogether, one that not only binds us, but explains how we came together.

Rich Brueckner is President of insideHPC.com

Preface

Galaxies, Why Bother?

Why does the universe bother to create galaxies? Why not just create stars and have stars and their planets floating around in space, each stellar system unto itself? It turns out that galaxies are a very efficient way to form stars. This is especially the case for the flattened disk or spiral galaxies. In the later stages of their lives, stars shed their outer envelopes. This material is mixed into the interstellar medium within galaxies and is used to form new generations of stars. Over time a galaxy will evolve from having a smaller number of higher mass stars with shorter lives to many long-lived stars of lower masses.

Our universe is just under 14 billion years old. It was born in a Big Bang, and as it cooled down, very small density fluctuations grew under their own gravity to form the first galaxies and stars before the universe was even 500 million years old. After the first few minutes of the Big Bang the universe was made of hydrogen and helium and the mysterious dark matter that dominates the self-gravity of galaxies. The other elements have been formed in stellar interiors subsequently.

People and life as we know it require carbon, nitrogen, oxygen and many other elements necessary for biochemistry. These are formed in stellar interiors through thermonuclear fusion. Advanced life requires planets to live on, and stars to provide light and heat in support of life. Galaxies are the natural homes for large collections of stars. They are recycling factories that allow matter from old stars – enriched with higher levels of carbon and other heavier elements - to be redistributed and to form new generations of stars. Without this recycling process, our Sun and Solar System would never have formed some 4.6 billion years ago. This process keeps galaxies active for billions of years. Our own Milky Way galaxy is thought to have over 200 billion and maybe as many as 400 billion stars, some more massive than our Sun but many more which are less massive.

Galaxies needed to exist, at least our own Milky Way has to exist, as a prerequisite for ourselves!

Stars and galaxies change over their lifetimes, and in the beautiful images that follow we will see many different types of galaxies, including some in the process of merger with their neighbors. In very broad terms we classify galaxies into three types: spiral, elliptical, and irregular galaxies. Some spirals have prominent bars and are categorized as barred spirals.

The following illustration shows Edwin Hubble's classification scheme, developed 90 years ago, for the main well-structured types: spirals, barred spirals, and ellipticals. The ellipticals are on the left, with the E0 classification representing the roundest ellipticals, and the E7 the most flattened. Actual galaxy images are located next to the stylized images for the various types.

This so-called 'tuning fork diagram' shows spirals on the top, and barred spirals on the bottom branch. Both types have spiral arms, with the most tightly wound being classified as Sa, and the looser spirals as Sc. If there is a bar in the center as for the lower branch of the 'tuning fork', then the B letter is added, so SBa through SBc.

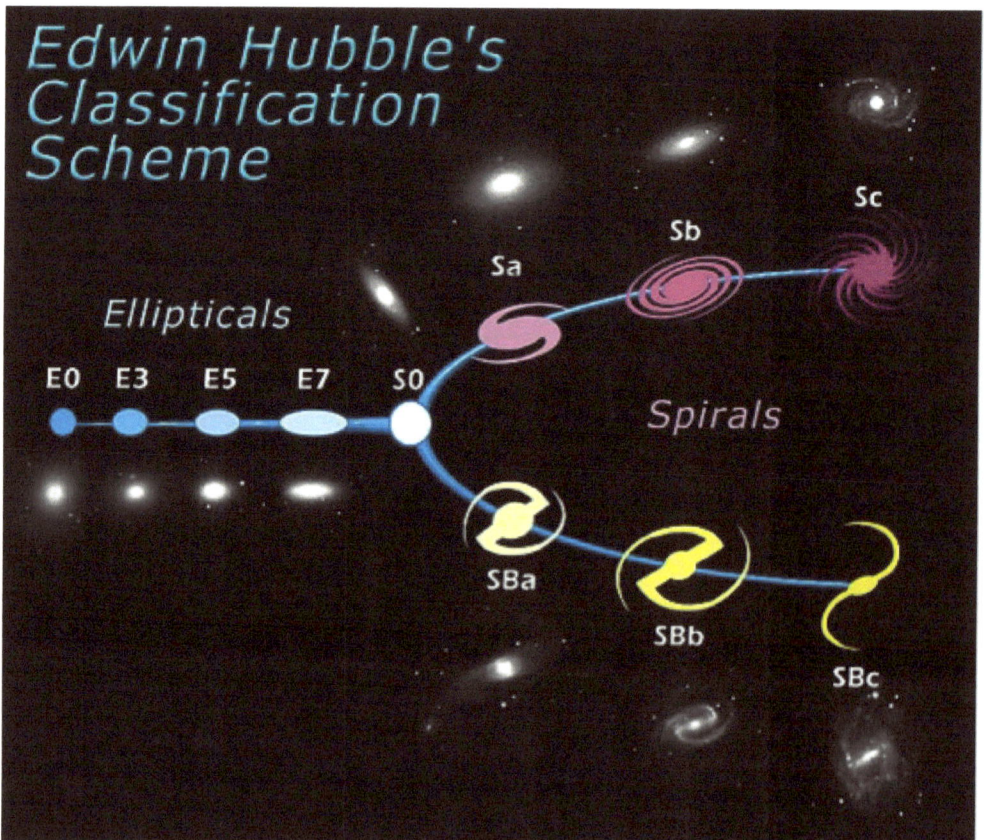

Edwin Hubble's 'tuning fork diagram' for galaxy classification.

This illustration was created for NASA by the Space Telescope Science Institute under Contract NAS5-26555, and is in the public domain.

Acknowledgements

I would like to thank Rich Brueckner for kindly agreeing to write the Foreword. It is a pleasure to acknowledge comments on the manuscript and visual appearance from Steve Campbell, Henry Fong, Prof. Prem Jain and Prof. Gregory Shields.

Bangkok
August, 2016

Chapter 1
"M" Galaxies are Nearby

The "M" galaxies are the Messier galaxies. These come from the famous Messier catalog that French astronomer Charles Messier initially published in 1771. Over 40 of the objects in the catalog are galaxies, while other members include star clusters, globular clusters and planetary nebulae. Since these were discovered long ago, they are some of the nearest of the known galaxies.

The most famous of the Messier galaxies: M31

M31 - Andromeda Galaxy

Our beautiful neighbor, M31, or the Andromeda Galaxy, is the best known galaxy apart from our Milky Way. M31 is a spiral galaxy, similar to, but larger than, our own galaxy. It has about two or three times the number of stars as the Milky Way.

M31 is visible to the naked eye on Moonless nights. Its existence has been known for over 1,000 years, but its nature as a galaxy separate from our own was not confirmed until around 1920, with the "Great Debate" between Harlow Shapley and Heber Curtis. Curtis was an advocate of many nebulae being external galaxies in their own right, and was proven correct.

Our galaxy and the Andromeda Galaxy are gravitationally bound together in our Local Group, along with about 50 other galaxies, which are almost all small dwarf galaxies.

The distance to the Andromeda Galaxy is around 2.5 million light-years (thus we are seeing it as it was 2.5 million years ago). The mass is estimated at over a trillion solar masses, and it has a star count of approximately a trillion stars. It is moving toward us at 300 kilometers per second (1/1000 of the speed of light). It is expected that M31 and our Milky Way will collide and merge in 4 billion years. (A video of a simulated collision can be found at this site: https://en.wikipedia.org/wiki/Andromeda–Milky_Way_collision .

Detailed studies of Andromeda indicate that it itself formed from a collision of two smaller galaxies more than 5 billion years ago.

M63 - Sunflower Galaxy

M63 is an Sb spiral in the direction of Canes Venatici. It was discovered in 1779 by Pierre Mechain, a friend of Charles Messier, who composed the Messier catalog. The nickname of this galaxy, the "Sunflower" galaxy is reasonably obvious, since it has a shape and coloration similar to a sunflower and exhibits beautiful yellow colors as well as blue. The latter are regions with recent ongoing star formation. Its spiral arms were first noticed in the mid-19th century.

The Sunflower galaxy is 37 million light-years distant, and is part of a group of galaxies together with M51.

M64 - Blackeye Galaxy

Image credit: NASA and The Hubble Heritage Team (AURA/STScI)

A collision of two galaxies caused the unusual appearance of the Blackeye Galaxy. The black area in this galaxy is due to massive amounts of dust on this side of the center.

While the stars are all revolving around the galaxy's center in the same sense, a significant amount of gas in the outer regions is moving in the opposite sense. From this, astronomers deduce that a smaller galaxy collided with M64 in the past, perhaps a billion years ago or more.

There is a boundary where the outer region gas, moving in the opposite sense of the stars, meets gas in the inner region, which moves with the stars. In the boundary region the gas clouds moving in opposite directions are colliding, leading to higher density regions, and new star formation is thus enhanced.

The pink regions are hydrogen gas glowing in the red part of the spectrum after absorbing ultraviolet light from new hot stars.

M81 - Bode's Galaxy

Image credit: NASA/JPL-Caltech - Spitzer Space Telescope

M81 can be seen with the naked eye by a skilled observer, since it is one of the nearest galaxies, just beyond our Local Group. It is located at a distance of 12 million light-years, and is in the constellation Ursa Major. When we say this, we really mean *behind* the constellation Ursa Major, since the constellations are all foreground collections of stars within our own galaxies. The apparent brightness of M81 is seventh magnitude (in the magnitude scale a higher number is fainter and each additional magnitude indicates a factor of about 2.5 times in brightness).

This image was obtained from a major orbiting infrared observatory, the Spitzer Space Telescope, and both near and far infrared data is represented. The near infrared is in the blue and far infrared (longer wavelength) is shown in red. Infrared radiation is produced by cool (red) stars, by dust, and by molecular clouds (cold clouds of gas containing molecular hydrogen and other molecules, even organic molecules).

M82 - Cigar Galaxy

Image credit: NASA/JPL-Caltech/STScI/CXC/UofA/ESA/AURA/JHU

M82 is an irregular, highly active, starburst galaxy. M82 is located in the direction of the constellation Ursa Major (the Big Bear). This image is a composite of data obtained from X-ray, optical and infrared observations.

The blue colors are the X-ray intensity from the Chandra X-ray Observatory. The orange and green colors are the optical data from the Hubble Space Telescope. The red colors are the infrared data from the Spitzer Space Telescope.

M82 is relatively nearby, about 12 million light-years. The image of the core, shown in the blue X-ray only inset image, is about 6,000 light-years across.

There are two bright X-ray sources which appear to be intermediate-mass black holes, in-between the stellar mass type and the supermassive type (millions of solar masses or more).

A bright supernova within M82 was detected in January 2014, and was the brightest Type 1a supernova seen in the last 40 years.

One of the black holes is located around 300 light years from the center of M82 and has a mass in the range of 12,000 to 43,000 times the mass of the Sun. The second black hole is 600 light years out from the center and has a small mass of over 200 but less than 800 times the mass of the Sun.

M83 - Southern Pinwheel Galaxy

Image credits: NASA, ESA, and the Hubble Heritage Team (STScI/AURA)

What makes this galaxy so beautiful? For me it is the open spiral arms gracefully trailing behind, and the Christmas tree light effect of the ruby red emission regions seen along the arms. Spiral arms are density waves, or concentrations, with higher density for gas and stars and formation of new stars. The red and pink colors so vividly presented in the lower image are due to gaseous nebulae and star forming regions - notice how they are associated with the spiral arms.

M83, or the Southern Pinwheel Galaxy or "Thousand Ruby Galaxy" is a barred spiral galaxy in the direction of the constellation Hydra, and 12 million light-years distant. Like all galaxies, it is behind the constellation, since constellations are apparent patterns of stars within our own galaxy.

So when we say that "it is in Hydra" as we see it in the night sky, we really mean "it is behind Hydra".

Also associated with spiral arms are dark, blackish areas. These are dust lanes. Dust is just that, small particles that include "metals" - which in astronomer-speak also refers to carbon, nitrogen, oxygen, silicon, etc. The dust is cold, usually less than 100 degrees absolute, and it absorbs optical and ultraviolet light and reradiates in the infrared part of the spectrum.

M87 - Virgo A

Image credit: NASA/STSCi (Hubble Space Telescope)

Note the almost spherical appearance, and complete absence of spiral arms or dust lanes. Also very noticeable is the jet extending toward the upper right of the image. This jet is prominent at radio frequencies as well.

It is believed to be powered by an accretion disk around a supermassive black hole in the center of M87. The matter in the jet is moving at highly relativistic speeds, that is, at a significant fraction of the speed of light.

M87 is a peculiar galaxy. It is also the central galaxy in the Virgo cluster, a rich cluster of galaxies. It is classified as a supergiant elliptical galaxy, and is one of the most massive of relatively nearby galaxies. It has over 10,000 globular clusters, around 100 times as many as our Milky Way.

M87 emits gamma rays, the most energetic type of light, and that presumably comes from the supermassive black hole vicinity.

While the galaxy is somewhat plain, the jet is extraordinarily beautiful in appearance.

M96

Image credit: "NGC 3368 ESO" by ESO/Oleg Maliy, CC BY 3.0

M96 has very open, loosely defined spiral arms. Gas and dust have an asymmetrical distribution, on the near side of the galaxy, and extending in toward the center. Notice that the compact glowing core is off-center.

M96 is also known as NGC 3368. This image is from the Very Large Telescope at the European Southern Observatory.

Its size is similar to our Milky Way, around 100,000 light years in diameter. It belongs to a group of galaxies known as the Leo I group, and is the largest member. M96 is 31 million light-years away, and was discovered in 1781. There is a black hole in the center with uncertain mass, but believed to be over 1 million solar masses.

This galaxy is found in the direction of the constellation Leo the Lion. Quite a number of background galaxies are visible behind M96. One is a large edge-on spiral in the upper left portion of the image (behind M96's outer spiral arm).

M100

M100 (also NGC 4321) is an exceptionally beautiful spiral galaxy, relatively nearby at 55 million light-years distance.

It is located in the Virgo Cluster, where it is one of the largest galaxies. We are looking "face-on" at the central region in this photo from the Hubble Space Telescope. Notice how bright regions with new star formation are located along the beautifully detailed spiral arms. Many dust lanes are seen as dark bands permeating the spiral arms.

M100 is about the same size as the Milky Way, with a diameter of 107,000 light-years. It was discovered in 1781.

There is a small bar in the center with a radius of 3,000 light-years and with enhanced star formation. In 2006 a supernova was discovered in M100.

M104 - Sombrero Galaxy

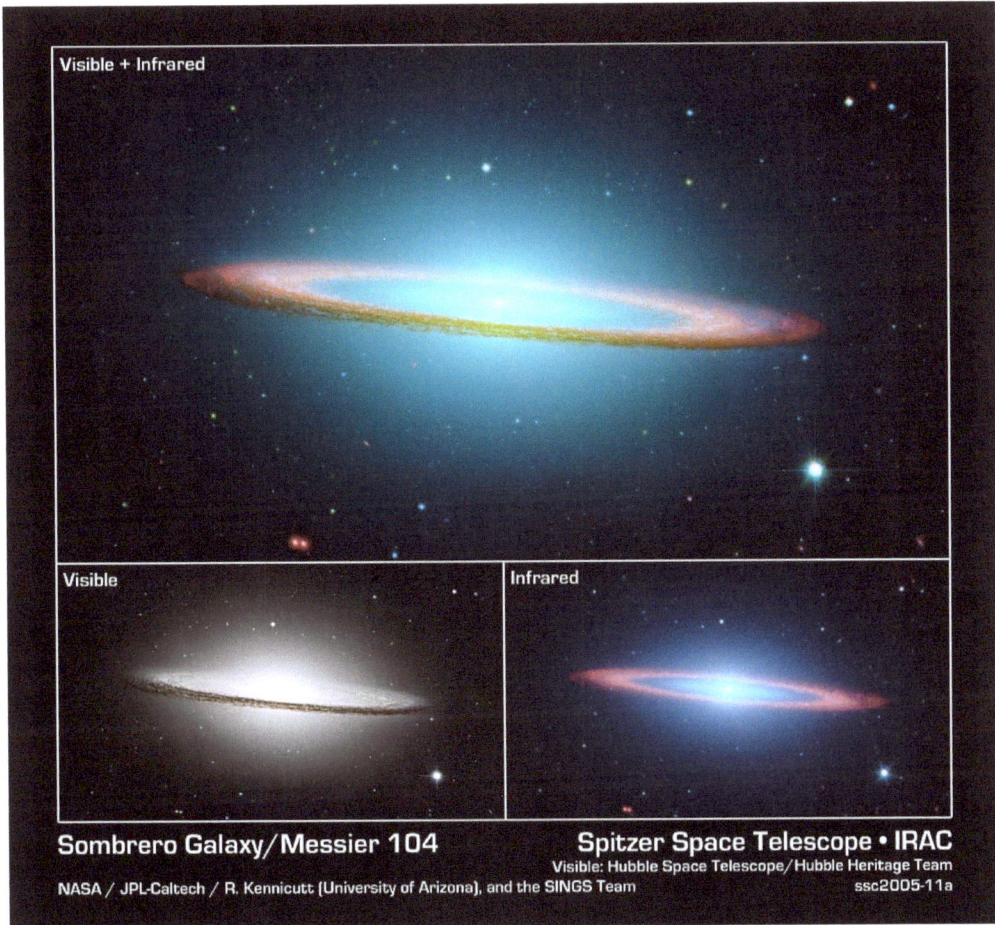

Image credit: NASA / JPL-Caltech / R. Kennicutt (Univ. of Arizona) and the SINGS Team

M104 has a large bulge which is clearly visible as we are seeing the galaxy almost edge on. It has tightly wound spiral arms and very prominent dust lanes in the disk (lower left image). This gorgeous image is a composite of the infrared and visible emission, with the dust regions and spiral arms clearly seen in the infrared.

The infrared image in the lower right is from the Spitzer Space Telescope, while the visible is from the Hubble Space Telescope.

The Sombrero Galaxy was discovered in 1781 and is 50,000 light-years in diameter, about half the diameter of our Milky Way. M104 is one of the brightest nearby galaxies at 9th magnitude (apparent), and is 28 million light-years away, in the direction of the constellation Virgo. It is located in a filament of galaxies at the south end of the Virgo cluster. It is large in the sky, with an apparent size that is one quarter of the Moon's.

M104 contains a supermassive black hole of around 1 billion solar masses, which is very much more massive than our Milky Way's central black hole. It also contains a large number of globular clusters (very dense spherically shaped clusters with hundreds of thousands of stars).

Large Magellanic Cloud

Image credit: NASA (C141 flight)

The Large Magellanic Cloud (LMC) is a dwarf barred spiral galaxy that has been disrupted by the Milky Way's gravity. This companion galaxy of our Milky Way is located in the Southern Hemisphere, is easily visible to the naked eye, and is at a distance of around 163,000 light-years away. It is 14,000 light-years in diameter.

We include the Large and Small Magellanic Clouds and some nearby dwarf galaxies in this chapter. Although they are not in the Messier catalog, because they are too faint, or too far south for Messier to have catalogued, they are also in our immediate galactic neighborhood.

The estimated mass of the LMC is 10 billion solar masses, around one percent that of the Milky Way. Nevertheless, it is the fourth largest galaxy in the Local Group of galaxies (the three largest are Andromeda, the Milky Way, and the Triangulum galaxy). It is not clear that the LMC is actually orbiting the Milky Way, but it is clearly within the range of tidal influence. The LMC has active star formation; its Tarantula Nebula is the most active star-formation region within the Local Group.

The Large Magellanic Cloud had its first recorded mention in 964 AD by a well-known Persian astronomer Al Sufi. It is located in the constellations Dorado and Mensa.

Small Magellanic Cloud

The Small Magellanic Cloud (SMC) is a dwarf irregular galaxy, and like its sister, the LMC, it appears to have a bar that has been disrupted. It is naked-eye visible as a faint hazy object extending across three degrees of sky. Its linear extent is 7,000 light-years, about half that of the LMC, and its mass is about 7 billion solar masses, also less than the LMC.

The SMC is further away than the LMC, at a distance of 200,000 light-years. Like the LMC, it is located in the Southern Hemisphere, around 20 degrees away from the Large Magellanic Cloud, in the constellations of Tucana and Hydrus.

There is a bridge of gas which connects the Small Magellanic Cloud to the Large Magellanic Cloud; this is evidence of a tidal interaction between the two.

Italian explorers had spotted the two Clouds in the late 15th century. Nevertheless, they are both named for Ferdinand Magellan, the Portuguese explorer, whose shipmate rediscovered them during Magellan's circumnavigation of the globe.

Sculptor Dwarf

Image credit: European Southern Observatory

Because it is so faint and diffuse, the Sculptor Dwarf galaxy was only discovered relatively recently, in 1937. It is believed to be older than the Milky Way, which makes it very useful in studying the earliest epochs of star formation.

There are a number of nearby galaxies which are not in the Messier catalog because of their low masses, luminosities, and surface brightnesses. These are known as dwarf galaxies.

The Sculptor Dwarf Galaxy is a neighbor of our Milky Way, and is one of over a dozen known satellite galaxies orbiting our own. This galaxy is about 280,000 light-years distance, in the constellation Sculptor.

UCG 8201

Image credit: NASA/ESA Hubble Space Telescope

The galaxy UGC 8201 has many relatively young stars due to a several hundred million-year episode of star formation, which has now ended.

This dwarf irregular galaxy is just 15 million light-years away in the Draco constellation. It is gravitationally bound to the much larger M81 galaxy.

Dwarf galaxies are hard to detect and measure due to their low intrinsic luminosity. But they are important to study. In particular, some dwarf galaxies appear to have higher than average ratios of the elusive dark matter as compared to ordinary luminous matter.

PGC 51017

This is a beautiful blue (azure) compact dwarf galaxy; it is blue because it has very active star formation. The galaxy has recently formed many young, hot, massive, blue stars which determine the overall color. It is a dwarf irregular galaxy, and includes a number of active star forming clusters within the galaxy.

Chapter 2
"N" Galaxies are Far Away

The New General Catalogue was compiled in 1888 by John Louis Emil Dreyer. This was about 100 years after the Messier catalog, and it contains 7,840 objects, some of which are galaxies. An additional 5,386 objects were added in the Index Catalogues. Galaxies in these catalogs are referred to as NGC or IC galaxies. They are generally substantially farther away than M galaxies. Some of the most beautiful are shown in this chapter.

NGC 613

Image credit: ESA/Hubble & NASA and S. Smartt (Queen's University Belfast)

In this gorgeous image, dark dust lanes provide a wonderful counterpoint to the bright blue areas of intense star formation. The core of the galaxy is white as a result of the heavy concentration of stars in the center. A huge black hole is at the very center and has a mass of some tens of millions of stars. It continues to grow larger, consuming stars, gas and dust and the accretion disk around the black hole is a strong radio source. However the black hole has not been detected in optical and infrared wavelengths.

NGC 613 is a barred spiral galaxy located in the direction of the constellation Sculptor at a distance of some 65 million light-years. When the light left this galaxy, the dinosaurs were at the point of extinction here on Earth. NGC 613 was first noted by William Herschel in 1798.

NGC 986

Image credit: NASA/ESA Hubble Space Telescope

The bright blue stars and star clusters are young stars found in this most lovely galaxy, whose appearance is enhanced by a number of dark dust lanes. This photograph is from the Wide Field and Planetary Camera 2 on the Hubble Space Telescope.

NGC 986 is a spiral galaxy discovered in 1828 by James Dunlop. It is around 56 million light-years to this beautiful galaxy.

NGC 1097

Galaxy NGC 1097

NASA / JPL-Caltech / The SINGS team (SSC/Caltech)

Spitzer Space Telescope • IRAC

ssc2009-14a

Image credit: NASA / JPL-Caltech / The SINGS team

Due to the bright red coloring of the image, this galaxy has the appearance of something out of a science fiction movie. The image was obtained in the infrared from the Spitzer Space Telescope. Here the image is rendered with colors visible to the human eye.

This galaxy has a very prominent bar, and at the center, the "eye" is a ring of stars surrounding the black hole. As material flows in, toward the central bar, it promotes new star formation.

NGC 1097 contains a supermassive black hole, with a mass of some 100 million times as large as the sun. NGC 1097 is 50 million light-years away.

NGC 1291

Image credit: NASA/JPL-Caltech

This is also an infrared image, from NASA's Spitzer Space Telescope. The stunning outer ring, shown in red, has very active star formation, while older stars are found in the central blue area.

The galaxy has an S-shaped bar in the center. NGC 1291 is ancient, around 12 billion years old, and when it was young, the stellar bar helped to drive gas toward the center.

Star formation has now shifted to the outer regions as the central region has run out of gas for new star formation.

The resulting image borders on the hypnotic. Let's move on to an even more incredible image.

NGC 2936 and 2937 - Arp 142 - the Penguin Pair

Interacting Galaxies Arp 142

NASA, ESA, the Hubble Heritage Team (STScI/AURA)
Hubble Space Telescope • WFC3/UVIS • STScI-PRC13-23a

Hubble Heritage

Image credit: NASA, ESA, the Hubble Heritage Team (STScI/AURA)

The larger of the two galaxies is NGC 2936, a spiral, and it is indeed shaped like a penguin. The companion, the elliptical galaxy NGC 2937, looks somewhat like an egg to which the penguin is lovingly attending. The galaxy pair is also found as object 142 in Arp's catalog of peculiar galaxies.

The larger spiral has very active star formation, as indicated by the blue colors, and prominent dark dust lanes. It is also significantly disturbed due to gravitational interaction with its companion galaxy. It appears that it was once a flat spiral galaxy, but it is now greatly warped and the spiral arms severely disrupted, as a result of gravitational tidal interactions with its elliptical neighbor.

The elliptical galaxy (the penguin's egg), may be larger in mass, judging by its gravitational effects. It is also older in terms of its star formation, and thus we do not see active star formation regions, or dust.

The galaxy at the top of the image is an unrelated foreground galaxy, either a spiral seen edge-on, or an irregular galaxy.

NGC 3079

Image credit: NASA, Gerald Cecil (University of North Carolina), Sylvain Veilleux (University of Maryland), Joss Bland-Hawthorn (Anglo-Australian Observatory), and Alex Filippenko (University of California at Berkeley).

The gorgeous galaxy NGC 3079 contains many reddish areas that are regions of ionized hydrogen plasma. There is a central active region with a bubble of hot gas over 3,000 light-years in extent. The gas bubble is due to high-speed particles and indicates a high rate of star formation (a 'starburst') in the galaxy's core. The outflow appears to be around a million years old.

The gas is expected to return to the galaxy's disk in the future and trigger additional star formation. The disk is some 70,000 light-years in diameter.

NGC 3521 - the Bubble Galaxy

Image credit: ESA/Hubble & NASA and S. Smartt (Queen's University Belfast)

The Bubble Galaxy is a spiral belonging to a class possessing fluffy appearance and known as 'flocculent spirals'. While there are clearly spiral arms in NGC 3521, they are not as well defined as in other spiral galaxies. Such fluffy-looking galaxies are more common than the grand design spirals.

The bright star formation regions tend to line up with the spiral arms, but perhaps not as clearly as in 'grand design' spirals. The contrast of density of matter is not as great in these galaxies. The relatively small density fluctuations result in a fluffy appearance not unlike stratocumulus clouds. The Bubble galaxy is at a distance of around 40 million light-years in the constellation Leo, and was discovered by William Herschel in 1784.

NGC 3921

Image credit: ESA/Hubble & NASA

NGC 3921 shows clear signs of major disruption. As you examine it more closely, you can see that it is a pair of merging galaxies. Both are spiral-type galaxies with disks. There has been a great deal of gravitational dislocation of the spiral arms as these galaxies merge, resulting in loops and streams of stars and gas.

Astronomers estimate that the collision has been going on for 700 million years, and that the two merging galaxies are of approximately equal masses.

The merger enhances star formation rates, and observations indicate over 1,000 bright star forming regions in this "new" galaxy. The result is lovely blue colors due to the young stellar population.

The Antennae Galaxies - NGC 4038 and NGC 4039

Image Credit: Hubble/European Space Agency

This is a pair of colliding galaxies, presenting an absolutely stunning image. Each of the two spiral galaxies is not so different from our Milky Way. They have spent hundreds of millions of years in a violent cosmic dance. Stars are being ripped from the respective galaxies and flung out into an arc between the two.

Gas clouds are in reddish colors, and the star-forming regions are seen in blue. The Antennae are in starburst mode, which is a phase of very rapid star formation. There are also many prominent dust lanes (dark areas).

This is a composite image based on both visible and near-infrared data from the Hubble Space Telescope.

NGC 4725

Image credit: NASA/JPL-Caltech/SST/R. Kennicutt (University of Arizona) and the SINGS Team

One sees in this infrared image very broad spiral arms, almost circular in appearance, with active star formation and extensive dust lanes. The location and appearance of the arms is similar to what is seen in its optical images. There is a very bright center.

This galaxy is a Seyfert galaxy, with an active galactic nucleus and presumably a supermassive black hole at the center. The resulting image looks like a cosmic eye.

This image was acquired by the infrared Spitzer Space Telescope and is rendered in false-color. NGC 4725 is a barred spiral at a distance of 40 million light-years, in the constellation Coma Berenices.

NGC 4845

Image credit: ESA/Hubble & NASA and S. Smartt (Queen's University Belfast)

This galaxy is quite gorgeous, as a result of very prominent foreground dust lanes, as well as the bright center. This is a spiral galaxy with a bright galactic bulge in the center. There is believed to be a supermassive black hole in the center of NGC 4845. By looking at the motions of the stars in the very center, which move around faster than would otherwise be the case, the black hole is estimated to be three hundred thousand times as massive as the Sun.

An X-ray flare was detected in 2013 from the black hole swallowing matter. The estimated mass of the object was much greater than Jupiter's mass, so it was likely either a very large planet, or a small brown dwarf star.

NGC 4845 is located in the direction of Virgo, has a redshift of z = .0041 (see Glossary, redshift) and a distance of 47 million light-years. It is of type Sab and was originally discovered in 1786 by William Herschel.

NGC 4911

Image credit: NASA/ESA/Space Science Institute

NGC 4911 is a beautiful face-on spiral galaxy, located in the Coma cluster, at a distance of 320 million light-years, as captured by the Hubble Space Telescope. The hydrogen gas clouds within NGC 4911 are seen in a pinkish hue, and are the birthplaces for new stars.

This is one of over 1,000 galaxies that are gravitationally bound to one another within the Coma cluster. The cluster is dominated by mysterious, invisible dark matter, as are all galaxies and all galaxy clusters. There is a large amount of very hot gas in the intracluster medium between the galaxies, that emits X-rays. Indeed, the amount of ordinary matter in hot gas located between the galaxies exceeds the amount of ordinary (not dark) matter found in all the galaxies within the cluster.

Clusters are composed of about 90% invisible dark matter, 9% hot gas (visible in X-rays), and about 1% stars and gas within the galaxies themselves.

NGC 6240

Image credit: NASA, ESA, the Hubble Heritage (STScI/AURA)-ESA/Hubble Collaboration, and A. Evans (University of Virginia, Charlottesville/NRAO/Stony Brook University)

What is happening in this galaxy!?

This highly irregular galaxy, named NGC 6240, is located 400 million light-years away. A NASA article describes it as "bizarrely-shaped" and as a "mess of gas, dust and stars [which] bears more than a passing resemblance to a butterfly and a lobster".

Perhaps they see the lobster in blue and the butterfly in pink. Or is it the other way around?

Two galaxies have collided and there are two supermassive black holes, which will eventually merge. At present they are only 3,000 light-years apart, which is just 1% of the size of the galaxy. Within a few million years the two black holes should merge into a single larger one, and in the process there will be an immense amount of energy produced in the center of this monstrosity. But what a lovely monstrosity it is!

NGC 6814

Image credit: NASA, GALEX

This is an ultraviolet image from the **GALEX** Galaxy Explorer space-borne telescope. One sees a bright center. The bluer regions have young stars and the spiral arms are visible. The effect is rather like an impressionist era oil painting due to the low-resolution nature of the image.

This is an energetic galaxy, a Seyfert galaxy, with an active nucleus. X-rays and ultraviolet emission have been detected in this central region.

NGC 6814 was discovered by William Herschel in 1788. Of course he did not know that these were very distant galaxies, he just knew they were nebulae, presumably within our own galaxy. It is located in the constellation Aquila.

NGC 6861

Image credit: ESA/Hubble & NASA

This exceptionally beautiful galaxy was discovered in 1826 by James Dunlop. It has very prominent dust lanes, seen as dark bands that circle the bright center. Its unusual appearance may be the result of a relatively recent merger. NGC 6861 is classified as a lenticular galaxy, intermediate between the spiral and elliptical classifications. It is a member of a small group known as the Telescopium Group.

NGC 6872 - the Condor Galaxy

Image Credit: FORS Team, 8.2-meter VLT Antu, ESO; Processing & License: Judy Schmidt

This is one of the longest barred spiral galaxies known. It measures over 700,000 light years across, or about 7 times as wide as the Milky Way Galaxy. The very extended spiral arms are thought to be a result of an ongoing collision with its neighbor, the galaxy IC 4970. That galaxy is visible just above NGC 6872 in the center of this image.

Note the bright blue star-forming regions, especially in the upper arm.

NGC 6872 is in the direction of the constellation Paco (the Peacock) but is 300 million light years more distant.

NGC 7252 - the Atoms-for-Peace galaxy

Image credit: Hubble Space Telescope, NASA and ESA

The peculiar appearance of this galaxy is thought to be due to a collision between two galaxies around a billion years ago. One sees streams of stars, gas and dust almost spinning out in various directions.

In the center are found many clusters of young bluish stars, and it is believed that the galaxy over time will develop toward an elliptical galaxy, as the gas supplying the formation of stars is used up.

NGC 7252 is a peculiar galaxy in Aquarius, and located 220 million light-years away. It resembles the stylized version of electron orbitals around an atomic nucleus and has been nicknamed the Atoms-for-Peace galaxy. ("Atoms-for-Peace" was a famous speech given in 1953 by U.S. President Dwight D. Eisenhower on nuclear nonproliferation and the peaceful use of atomic energy).

NGC 7714 - Arp 284

Image Credit: ESA, NASA, Acknowledgement: A. Gal-Yam (Weizmann Institute of Science)

NGC7714 appears to be a barred spiral galaxy, but has been significantly disrupted by the tidal interactions with its companion, which is outside of this frame (toward the left) and some 80,000 light-years away from NGC 7714. It is a starburst galaxy, with enhanced star formation induced by its companion's tidal effects, resulting in the beautiful blue shades we see here.

NGC 7714 and NGC 7715 (the companion) are a pair of interacting galaxies, located 100 million light-years from Earth. The galaxy pair is also found in Arp's catalog of peculiar galaxies as entry Arp 284. Only NGC 7714 is shown in this image. NGC 7714 was discovered in 1830 by John Herschel and lies in the constellation Pisces.

Chapter 3
"z" Galaxies are Incredibly Far Away

There are many, many other galaxy catalogs that have been produced in the 20th and 21st centuries; we do not have space here to go through even a partial list. This chapter has many images of very high redshift galaxies, billions of light-years distant, and referred to as 'z' galaxies here, since z is the symbol for redshift. The larger the z value, the farther away is the galaxy.

Image credit: Pablo Carlos Budassi, Creative Commons Attribution-Share Alike 3.0 Unported license. The observable universe represented logarithmically, our Solar System shown at center and the cosmic microwave background radiation from the Big Bang at the edge.

Hoag's Object

Image credit: NASA, The Hubble Heritage Team (STScI/AURA)

This unusual ring galaxy was discovered in 1950. Incredibly, inside the ring of Hoag's object, near the top, another ring galaxy is seen in this image from the Hubble Space Telescope! The ring's stars are young blue stars, and the stars in the central region of are yellow for the most part.

Ring-shaped galaxies may be formed through collision with another galaxy, and in this case it is thought that the encounter may have been 2 or 3 billion years ago. The blue ring could be the remains of the other galaxy.

Hoag's Object is relatively nearby, with a redshift of about z = 0.04. The galaxy is about 120,000 light-years across, around the size of our Milky Way Galaxy or a bit larger.

LO95 0313-192

Image credit: ESA/Hubble & NASA; Acknowledgement: Judy Schmid

This galaxy is seen edge on, and is the larger galaxy to the left hand side of the image. It has very prominent dust lanes and a bright central bulge. It also has intense jets emanating from the center, seen at radio frequencies. Jets are quite frequent in giant elliptical galaxies, but much less common with spirals. The dust lanes are quite prominent.

LO95 0313-192 is a billion light-years away, with a modestly large redshift of 0.067, and is in the direction of the constellation Eridanus.

The other spiral galaxy is a companion, and has the moniker J0131549.8-190623, an unattractive name for such a pretty galaxy.

3C 295

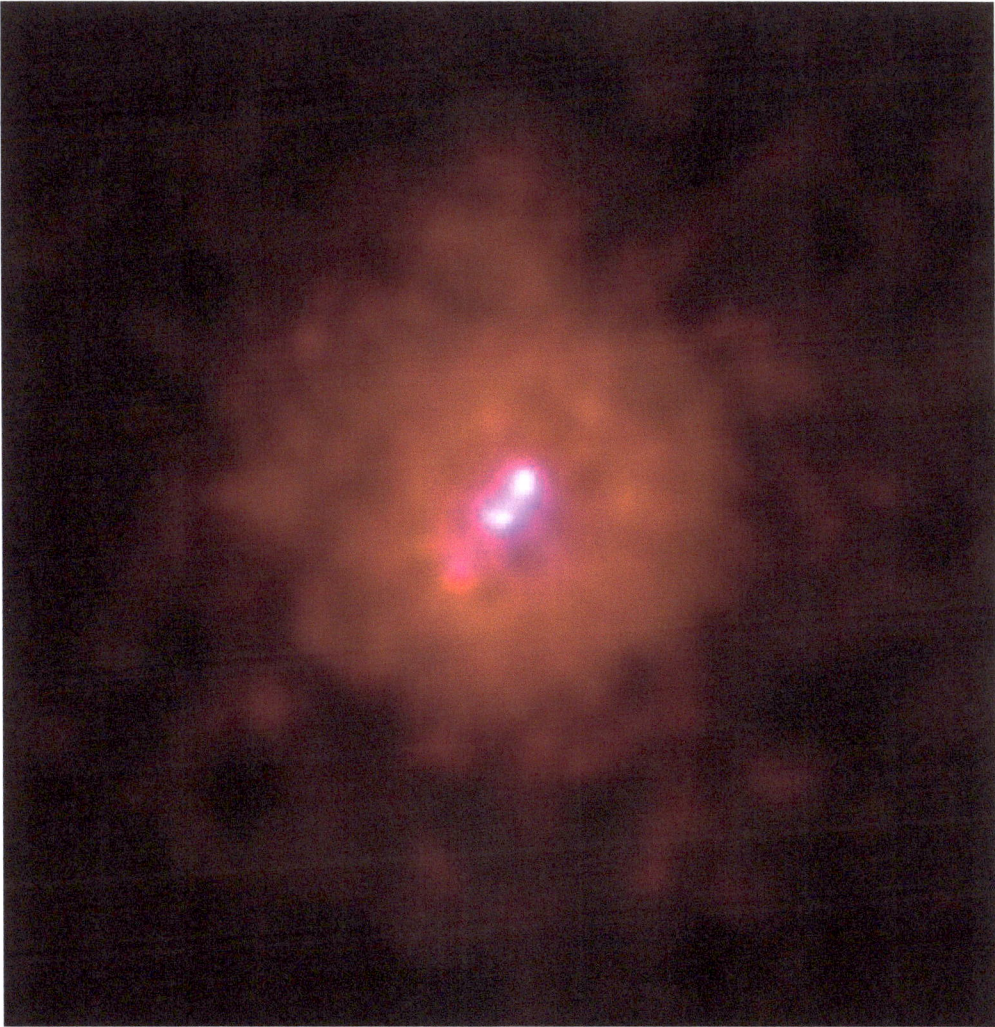

Image credit: CHANDRA X-ray Observatory, NASA/CXC/SAO

This is an X-ray image of a galaxy that was first discovered as a radio source, hence the 3C designation, from the Third Cambridge Catalogue of radio sources. At the time of its discovery in 1960 this was the most distant known extragalactic object, with a redshift of 0.46.

The radio emission (not shown) is from an elliptical galaxy, that is also a strong X-ray source. It turns out to be located near the center of a cluster of galaxies, so in this X-ray image one sees diffuse emission from hot cluster gas as well. The cluster contains hundreds of galaxies, whose gravitational potential is strong enough to heat the gas to millions of degrees.

3C 295 is located in the direction of the constellation Boötes.

3C 279

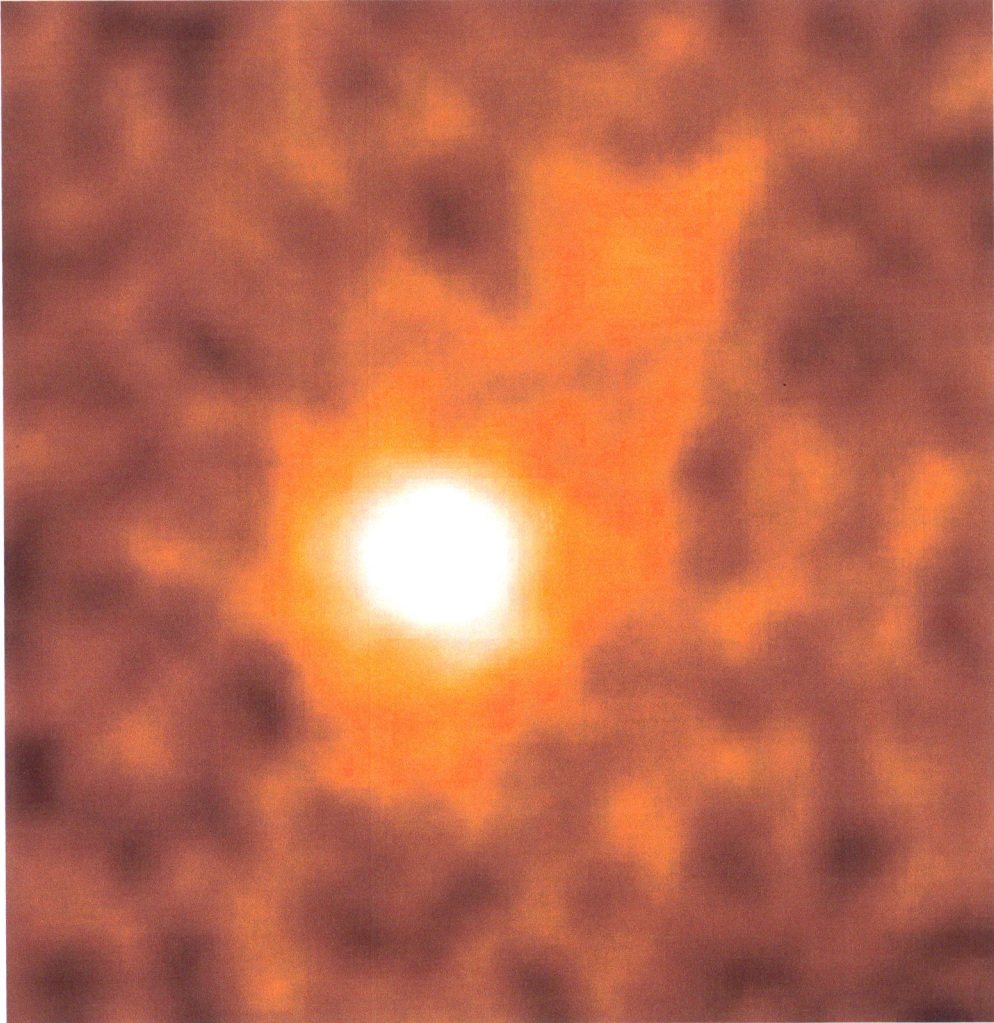

Image credit: EGRET team, Compton Observatory, NASA

What we are seeing in this image is the center of a galaxy located at a redshift of 0.54. It has a highly energetic nucleus that emits gamma radiation, as well as shining at radio and optical frequencies. This is one of the brightest known gamma ray objects, and it has been observed by the Compton Gamma Ray Observatory in 1991 and more recently by the Fermi Space Telescope.

3C 279 is labelled as an "optically violent variable quasar". The 3C designation indicates that the object was originally detected as a quasi-stellar radio source (quasar). The radiation from 3C 279 is highly variable in the optical band as well.

Dragonfly Galaxy

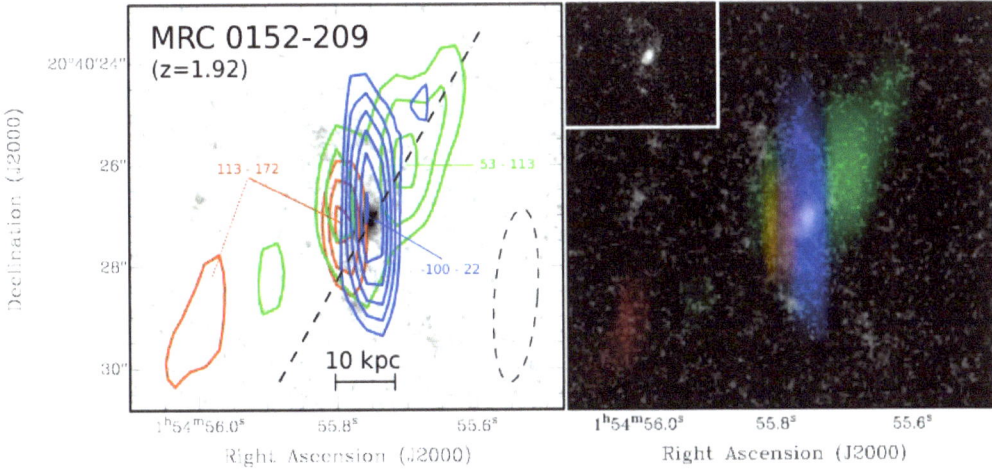

Image credit: https://inspirehep.net/record/1269695/files/MRC0152coloroverlay3.png;
Hubble Space Telescope

This galaxy contains very large molecular gas clouds, generally associated with significantly enhanced star formation. The contours on the left and the color enhancement shown in blue, red and green hues on the right are overlaid on the original Hubble Space Telescope image. They represent the strength of carbon monoxide emission as measured by the Australia Telescope Compact Array radio telescope. The three colors represent different clumps of gas as indicated by their velocities, and the overall scale is similar to that of our Milky Way galaxy.

Located at a redshift of $z = 1.92$, the Dragonfly Galaxy (also named MRC 0152-209) is a very bright galaxy at radio wavelengths and is the most infrared-luminous radio galaxy found at high redshift in the southern hemisphere sky.

The Cosmic Eyelash

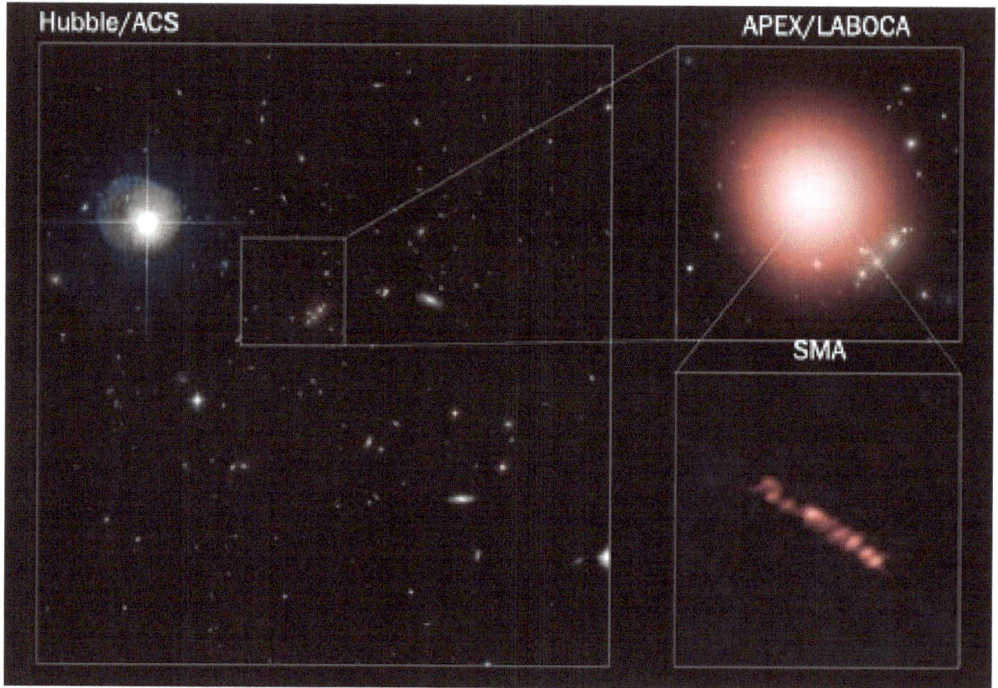

Image credit: M. Swinbank et al./Nature, ESO, APEX; NASA, ESA, SMA

The Cosmic Eyelash is located at a redshift of 2.33. It is more formally designated as SMM J2135-0102 where the SMM refers to submillimeter wavelengths in the far infrared. It is an Ultra-luminous infrared galaxy (ULIRG).

The image is from the Hubble Space Telescope, and the upper right inset is from a submilimeter bolometer array at the Atacama Observatory in Chile.

The lower right inset in the image is the far infrared detection from the SAO Submillimeter Array on Mauna Kea, Hawaii.

From the Herschel Space Observatory outflow of gas is detected. This gas fountain has winds flowing at 700 km/s; the gas is expected to fall back into the center since it is moving at less than escape velocity. Thus it should promote star formation at a later time.

Carbon monoxide (CO) is also detected, and the dust spectrum peaks around a wavelength of 350 microns.

The galaxy brightness is substantially magnified by gravitational lensing (see Glossary).

B3 0727+409

Image credits: X-ray: NASA/CXC/ISAS/A. Simionescu et al. Optical: DSS

This quasar (highly active galactic nucleus) is at a redshift of 2.5 and corresponds to an age of the universe of 2.7 billion years, some 11 billion years ago. There is a massively long jet, which has been discovered by the Chandra X-ray Observatory. The X-ray intensity of the jet is shown in the inset in blue colors.

The jet is some 300 thousand light-years long, two to three times as large as our Milky Way galaxy.

The jet is evidence of a supermassive black hole that generates very highly relativistic electrons. It is believed that the X-radiation is caused by cosmic microwave background photons being scattered to much higher energies by these highly energetic electrons.

BDF3299

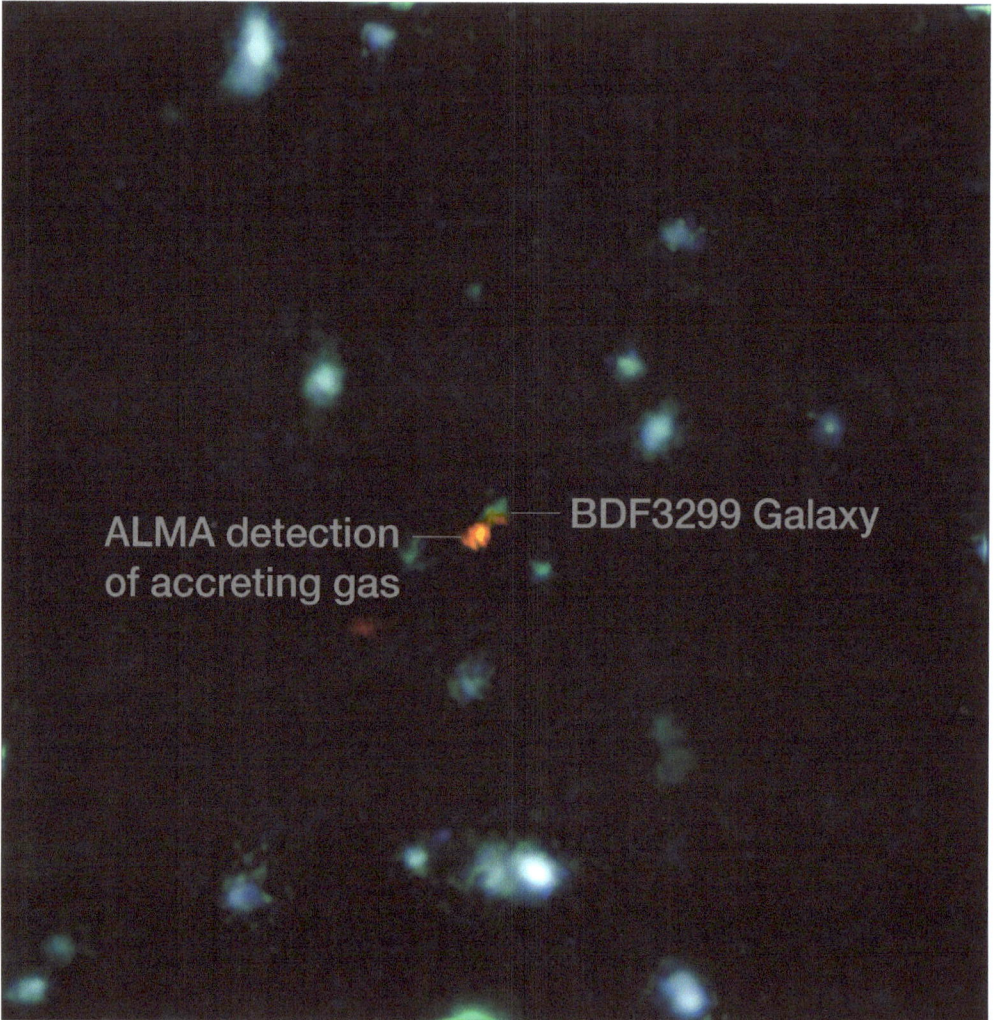

ALMA detection of accreting gas — BDF3299 Galaxy

Image credit: ESO/R. Maiolino

The extremely ancient galaxy BDF3299 is in the center of the image. It has a redshift of $z = 7$, so it is very old indeed. The light is from a time when the universe was only 800 million years old, some 13 billion years ago.

The red region is primordial gas falling into the galaxy as it undergoes formation. This is the most distant cloud of star-forming gas detected to date within the early universe.

The image is from ALMA, the Atacama Large Millimeter Array in Chile. ALMA is an array of 66 high-resolution and moveable antennas in the very dry Atacama desert, 5,000 meters above sea level. This millimeter or short-wavelength radio telescope can detect radiation from ionized carbon. The light from the galaxy is shifted by a factor of 1+z, or to 8 times longer wavelengths than the original emission, as a result of the expansion of the universe.

This discovery of ionized carbon is evidence of previous star formation, since carbon is not formed in the Big Bang, but only later in stellar interiors through thermonuclear fusion of helium atoms.

z8-GND-5296

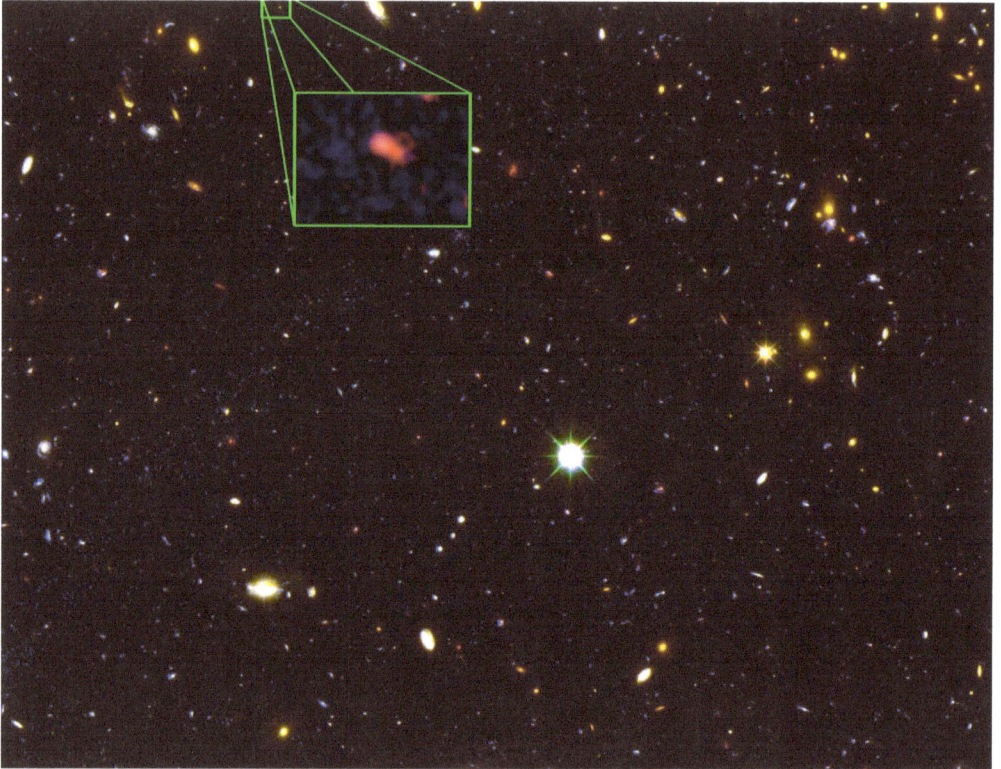

Image credit: V. Tilvi / S.L. Finkelstein / C. Papovich / A. Koekemoer / CANDELS / STScI / NASA.

This galaxy has a very high redshift of 7.51 and is one of the most distant galaxies ever discovered. It was detected in the infrared and its redshift measured from a hydrogen spectral line that normally would be seen in ultraviolet. However, due to the high redshift, the line is shifted all the way into the infrared. The spectroscopic observations were made from a major ground-based observatory, the W.M. Keck Observatory in Hawaii. The image shown here is a Hubble Space Telescope image and the insert shows a closeup. Note the very red color of this galaxy due to its large redshift and great distance.

It is forming stars at a rate of 300 solar masses per year, which is 100 times greater than the rate of star formation in our Milky Way galaxy.

EGSY8p7

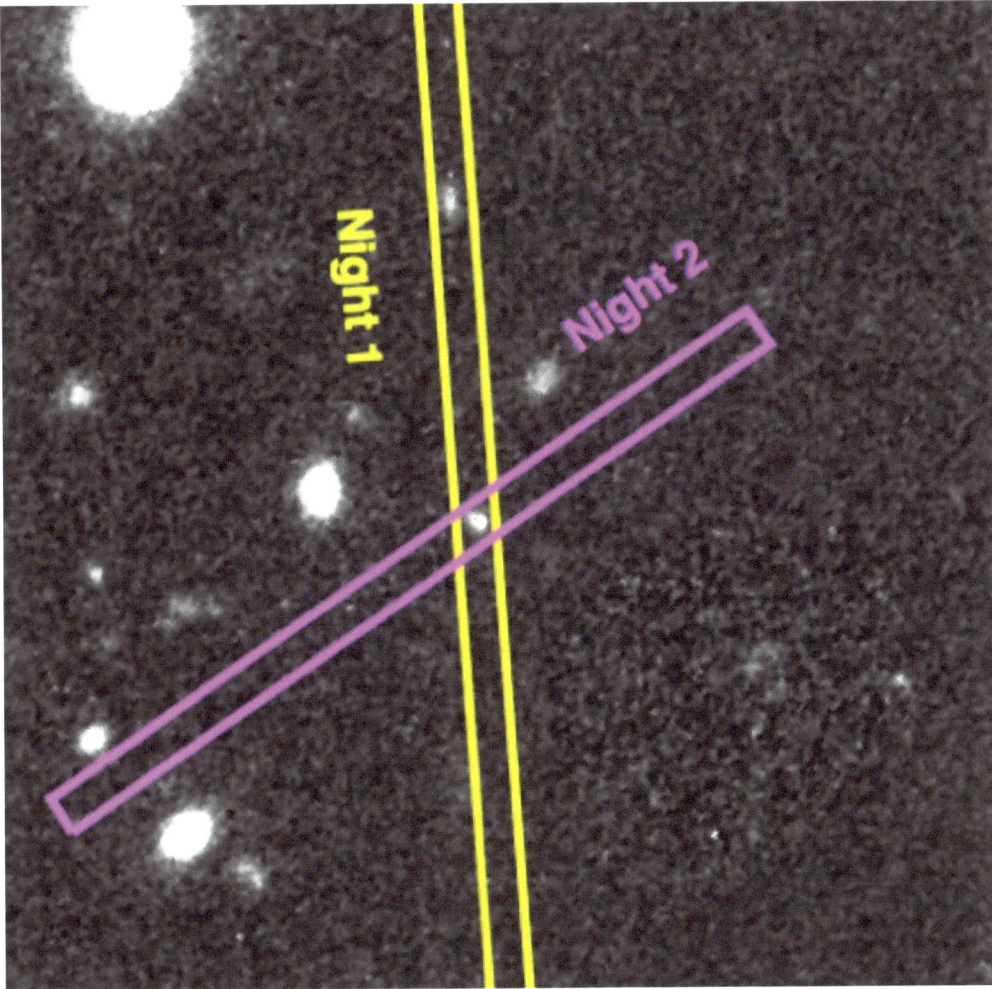

Image credit: Adi Zitrin, California Institute of Technology

EGSY8p7 was, until recently, the highest redshift galaxy known, with a redshift $z = 8.68$. The redshift was measured at the W. M. Keck Observatory, Mauna Kea, Hawaii by an international team from the U.S., the U.K. and the Netherlands.

The measurement is of a hydrogen line emitting in the ultraviolet in the rest frame of the galaxy, but shifting into the infrared; the wavelength increases by a factor of 9.68 due to the cosmological redshift.

The light comes from a time when the universe was less than 600 million years old; the light travel time is 13.2 billion years. It is very difficult to detect galaxies from within the universe's first billion years. The colors of EGSY8p7 suggest it is dominated by very hot stars, probably massive stars.

Looking so far back in time allows us to probe the period of cosmic reionization. The universe, after its first 400 hundred thousand years, had cooled off and the hydrogen and helium were then in the form of neutral atomic gas. But after 400 or 500 million years had passed some of the gas had already formed into stars and young galaxies. The young stars were quite massive and hot and thus emitted a large amount of ultraviolet radiation. This caused the intergalactic gas to heat up and become ionized again.

GN-z11

Image credit: NASA, ESA | P. Oesch (Yale University) | G. Brammer (STScI) | P. van Dokkum (Yale University) | G. Illingworth (University of California, Santa Cruz

As of March 2016, the current most distant galaxy known is GN-z11, since observations indicate that it has a redshift of over 11. It appears as a rather featureless red blob because of its great distance and high redshift. But it is beautiful in its own way because of what it tells us about early galaxy formation.

The light from GN-z11 is greatly shifted into the infrared due to the cosmic expansion. This high value of the redshift indicates that it is 13.4 billion light years away and had formed when the universe was only 400 million years old, or only 3% of its current age.

GNz-11 is a relatively low mass galaxy, with about 1 billion times the Sun's mass, and with a linear size 25 times smaller than our Milky Way galaxy. However, it is actively giving birth to new stars at a rate much higher than our Milky Way does.

This discovery has allowed astronomers to push back the frontier of when galaxies first formed. It is now thought that the very first stars formed when the universe was only 200 million years old.

Chapter 4

Groups of Galaxies and Clusters of Galaxies

Galaxies are not isolated entities.

They tend to clump together in groups and in clusters (clusters are very large groups). Our own Milky Way galaxy is a member of The Local Group, along with the Andromeda Galaxy and some 50 other galaxies. These groups and clusters are gravitationally bound systems. Their self-gravity is sufficiently strong to overcome the general expansion of the universe.

Galactic collisions and tidal interactions are frequent occurrences in groups; larger galaxies grow by cannibalizing their smaller neighbors. Star formation rates increase when galaxies merge.

HGC 31 - Hickson Compact Group 31

Image credit: NASA, ESA, S. Gallagher and J. English

Four galaxies are interacting in this group, with the bright central-left object consisting of two colliding dwarf galaxies. The scale in the photo of 50,000 light years is half the diameter of our Milky Way galaxy, so the separation between galaxies is relatively small.

The cigar-shaped galaxy at the upper left is connected to the double galaxy by a bridge of star clusters. The fourth member at the lower right is a disk galaxy, and a bridge of new star forming regions points to it.

Astronomers studying this compact group find many star clusters with new stars. Many of these have 100,000 stars or more, with ages of less than 10 million years. These recent star clusters have a bluish tint.

This image is a composite from three space-borne telescopes: the Hubble Space Telescope, the Spitzer Space Telescope, and GALEX (Galaxy Evolution Explorer).

Paul Hickson compiled the HGC catalog of galaxy groups in 1982. This group, HGC 31, is obviously going through a very active encounter phase, for the last several hundred million years, it is calculated. HGC 31 is at a distance of 166 million light-years.

HGC 59

Image credit: NASA, ESA/Hubble

One sees two large spirals in HGC59; one is face-on and more white in appearance and with broad spiral arms, and with a number of dust lanes clearly visible. The other, to the lower right, is inclined and more bluish. It appears to have a bar in the center and several spiral arms and accompanying dust and young bluish star regions are visible.

But the most beautiful image is a very irregular galaxy with hints of spiral structure and with many bluish regions full of young stars. Perhaps two galaxies collided to cause this irregular morphology.

Groups such as this often have galaxy-galaxy gravitational interactions. Other, more distant, background galaxies are also visible in the image. (The very bright image with four diffraction spikes is a star in our own Milky Way.)

This galaxy group is located in (behind) the constellation Leo. This image of HCG 59 is in the blue, yellow, and near-infrared portions of the electromagnetic spectrum.

HGC 62

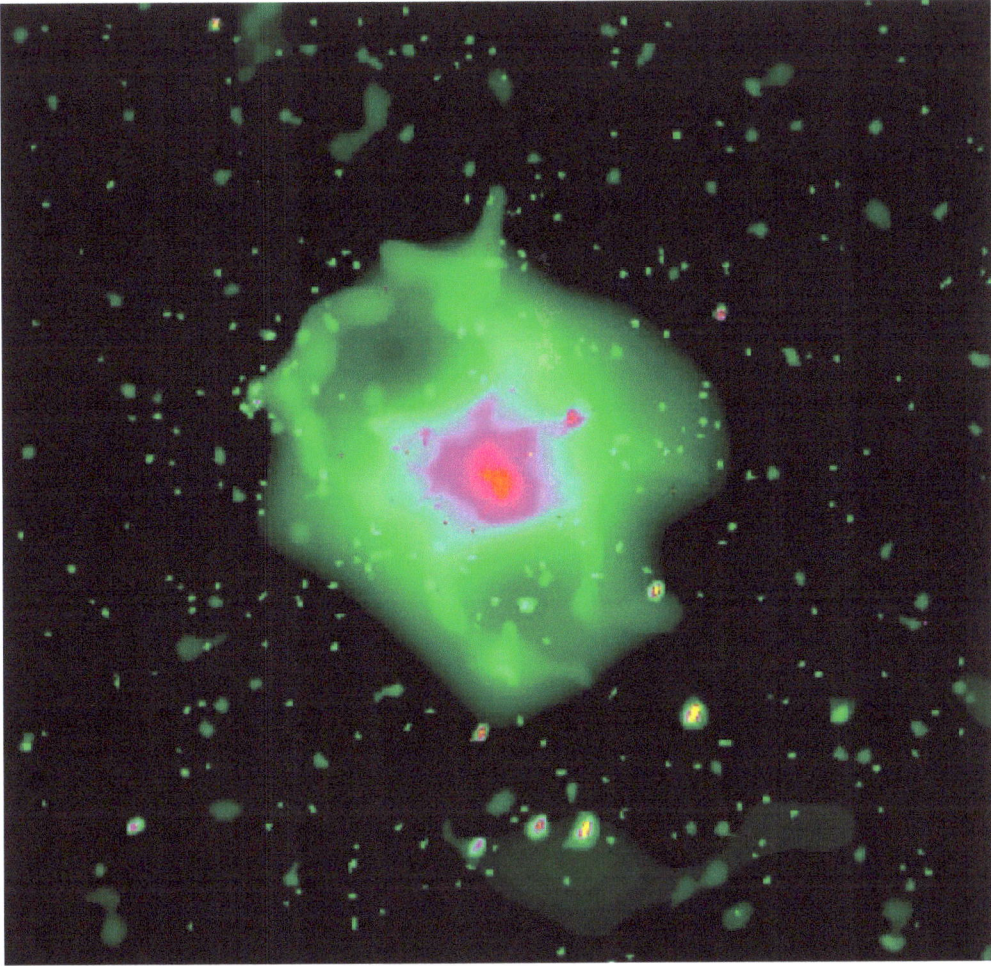

Image credit: NASA, CfA, J. Vrtilek et al.

There is evidence for a recent galactic merger in this group. HGC 62 lies in the Virgo constellation at a distance of 200 million light-years, and the elliptical galaxy NGC 4761 is near the center. Centrally based ellipticals are often found to have consumed other galaxies.

This image is from an X-ray telescope in orbit, the Chandra Observatory. What we see is the X-ray emission from very hot intergalactic gas between the galaxies in the group. This is a false-color image with red and purple colors indicating the highest X-ray intensity.

X-ray emission is associated with some groups of galaxies and virtually all rich clusters of galaxies. The gravitational potential is strong enough in such groups and clusters to heat the intergalactic gas (via frictional processes) to many millions of degrees. At such high temperatures the most efficient way for the gas to cool is via the emission of X-rays.

Seyfert's Sextet - HGC 79

Image credit: Hubble Space Telescope, NASA, ESA

The Seyfert Sextet is a group of *five* galaxies, four of which are at a distance of 190 million light-years. One member is not a true member of the group; it is at nearly 900 million light-years' distance. Which of the galaxies is more remote, can you tell?

And why is it called a sextet? Not because it is sexy, although it does look a bit that way. But because one of the "galaxies" in the group proper, the one that is to the lower rightmost, is actually an extension of its neighboring galaxy. In reality there are only four galaxies gravitationally bound together in this "sextet".

There is a beautiful dust lane in one of the galaxies that is seen edge-on. And there are lovely spiral arms in the smaller galaxy seen face on - but that's the *background* galaxy. It is not physically part of this group.

There is also a very beautiful blue spiral galaxy seen edge on at the top, oriented vertically, that is part of the group.

The group was discovered by Seyfert in 1951, and is also known as HGC 79 in the Hickson Compact Group catalog.

HGC 87

Image credit: Credit: Sally Hunsberger (Lowell Obs.), Jane Charlton (Penn State) et al.; Data: Hubble Legacy Archive, NASA/STScI; Processing: Robert Gendler

Here we see a large edge on spiral with extensive dust lanes, and on the right an elliptical galaxy. Another beautiful spiral is near the top of the image. The smaller spiral in the center is fainter and has a higher redshift, and is not physically associated with the three main members of the group.

This group of four galaxies is located around 400 million light-years away, in the constellation Capricornus. Three of the four main members are quite active, with active galactic nuclei and high star formation rates.

Stefan's Quintet

Image credit: NASA, ESA, and the Hubble SM4 ERO Team

One of the five galaxies in this very beautiful group is not physically part of the group. The spiral in the upper left is NGC 7320 and is much, much closer than the other galaxies at 40 million light-years. The remaining galaxies are at 290 million light-years distance. These other four galaxies are bound gravitationally with each other. Three of the galaxies in the group show clear signs of their mutual gravitational interactions - tidal effects. Their shapes are stretched, their spiral arms seem to be 'pulled out', and they show tidally instigated tails.

The largest central image is two galaxies, NGC 7318A and 7318B, and they seem to be literally dancing together. Many young blue stars are visible in the outstretched spiral arms and tails of these two.

(The images with four spikes each are foreground stars in our own galaxy.)

Stefan's Quintet is one of the most famous and beautiful of groups. Stefan's Quintet is also known as HCG 92, it was the first compact galaxy group to be discovered, in the 19th century.

Coma Cluster

Image credit: NASA/JPL-Caltech/GSFC/SDSS

The Coma Cluster has over 1,000 galaxy members, gravitationally bound to each other. It contains a large number of ellipticals especially in the central region, and two supergiant ellipticals dominate the center of the cluster. Spiral galaxies are found primarily in the outer regions of the cluster.

The Coma Cluster is perhaps the most famous of galaxy clusters, which have many more members than groups. It is located some 321 million light-years from Earth, in the constellation Coma Berenices.

This cluster was the first place that dark matter was discovered, by Fritz Zwicky in 1933. Dark matter is due to mysterious particles that pervade the universe, and we see it only due to its gravitational effects. Zwicky observed that the galaxies were moving around more rapidly than anticipated, at around 1000 kilometers per second rather than the 300 kilometers/sec expected due to the galaxy masses. The dark matter content is about 10 times the ordinary matter. The ordinary matter is itself distributed between the galaxies and hot gas.

The Coma cluster has been seen as an X-ray source since the 1960s, due to the very hot intracluster gas, which heats up in the strong gravitational potential of the cluster, especially because of the dark matter.

Virgo Cluster

Image credit: ESO / Chris Mihos and his colleagues using the Burrell Schmidt telescope shows the diffuse light between the galaxies belonging to the cluster. The dark spots indicate where bright foreground stars were removed from the image. CC BY 3.0

We have previously seen the galaxy M87, which is a predominant member of the Virgo Cluster. It is the largest galaxy in this image, seen at the lower left. It is found at the center of the cluster. There are other Messier galaxies in this image, M84 at right center, and M86 to its left. And one sees a chain running from the right side toward the upper center. The brightest member of the cluster is M49.

The cluster has roughly equal numbers of spirals and elliptical galaxies. The spirals are stretched out in a filamentary fashion. Our own Local Group is loosely connected to the Virgo Cluster as part of the Virgo Supercluster.

The Virgo cluster is located at 54 million light-years distance, in the constellation Virgo and consists of over 1,300 galaxies. Its mass is estimated at over 1,000 trillion solar masses. Virgo is also an X-ray source due to a substantial intracluster medium with very hot gas at 30 million degrees.

Abell 2744 - Pandora's Cluster

Image credit: NASA, ESA, J. Merten (Institute for Theoretical Astrophysics, Heidelberg/Astronomical Observatory of Bologna), and D. Coe (STScI)

The oldest and best known catalog of galaxy clusters is known as the Abell catalog. This cluster is Abell 2744, and is also known as Pandora's Cluster.

Abell 2744 is at a distance of 4 billion light-years. Thus we are looking back about 30% of the universe's lifetime when we observe this cluster. It's a rather exotic cluster, hence the nickname "Pandora's". It shows evidence of a collision and merger of four separate smaller clusters in the previous several hundred million years.

This image has been false-colored to show the distribution of hot gas that emits X-rays, and also the distribution of dark matter. The X-ray emitting regions are colored in red, and the dark matter is colored in blue. The cluster is also a radio source. And it has many orphaned stars which have been torn free from their parent galaxies and roam freely in the intracluster medium.

MACS J0416.1–2403

Image credit: NASA, ESA and the HST Frontier Fields team (STScI); X-ray image: NASA/CXC/SAP/G. Ogrean et al.

In addition to the cluster galaxies proper, this cluster has a number of background galaxies (behind the cluster) with distorted images. As their light passes through the cluster, their images are stretched, curved and magnified by the gravitational lensing effect, in accordance with Einstein's general relativity theory. This allows astronomers to determine the distribution of dark matter, which dominates the overall mass of the cluster MACS J0416.

The cluster is actually composed of two sub-clusters in the process of merging, and separated by less than 1 million light-years.

There is also a large amount of very hot gas heated in the strong gravitational field of the cluster; the smaller image at left is of the X-rays emitted from this hot gas. Based on the distribution of the X-ray emission and the dark matter distribution inferred from the optical image, the merger is in its early stages.

The galaxy cluster MACS J0416.1-2403 has a redshift of just under 0.4 and is located some 4.3 billion light-years away in the direction of the constellation Eridanus. The larger image is from the Hubble Space Telescope; this is one of six galaxy clusters studied as part of Hubble's Frontier Fields program.

MACS J0717

Image credit: X-ray: NASA/CXC/SAO/van Weeren et al.; Optical: NASA/STScI; Radio: NSF/NRAO/VLA

This galaxy cluster is extremely distorted, since it is the result of a collision between four smaller clusters. It is located at a distance of 5.4 billion light years.

This image combines the optical image from the Hubble Space Telescope with radio telescope data from the Very Large Array (pink) and Chandra X-ray Observatory data (blue diffused colors).

Large arcs of radio emission (pink) are due to shock waves resulting from collisions. The X-ray emission is highly clumped because of the four clusters which collided to result in this larger cluster.

MACS J0717 is one of the "Frontier Fields" project clusters of galaxies for the Hubble Telescope.

IDCS 1426

Image credit: NASA, ESA, and M. Brodwin (University of Missouri)

The galaxy cluster in this image is called IDCS J1426.5+3508 (IDCS 1426 for short). The image is actually a composite of 3 images, with diffuse X-ray emission as recorded by the Chandra X-ray Observatory shown in blue.

The X-ray emission is due to hot gas at a temperature of around 100 million degrees, found between galaxies and throughout the cluster core. The visible light observed by the Hubble Space Telescope is shown in green, and the infrared light from the Spitzer Space Telescope is shown in red.

Because the cluster is located around 10 billion light-years away, we see it at a time when the universe was less than one-third as old as it is now. The light has been redshifted significantly due to the universe's expansion during that time.

The cluster is very massive, one of the most massive known clusters at such a large distance. The cluster weighs in at 500 trillion solar masses, with 90% of its mass being in the form of the mysterious dark matter.

It was discovered in the infrared part of the spectrum by the Spitzer Space Telescope in 2012. It appears to have had a collision or merger with another cluster during the past half billion years, allowing it to reach its large size and mass.

El Gordo

Image credit: NASA, ESA, J. Jee (Univ. of California, Davis), J. Hughes (Rutgers Univ.), F. Menanteau (Rutgers Univ. & Univ. of Illinois, Urbana-Champaign), C. Sifon (Leiden Obs.), R. Mandelbum (Carnegie Mellon Univ.), L. Barrientos (Univ. Catolica de Chile), and K. Ng (Univ. of California, Davis)

Not as far away as the previous cluster, but even more massive, is El Gordo, also known as ACT-CL J0102-4915. It has been nicknamed El Gordo ("the fat one") because it is the most massive cluster known, at 3,000 trillion solar masses.

It is located around 7 billion light years away. El Gordo also emits the most X-rays of any cluster known. Most of the matter is in the form of dark matter, and in this composite image is shown in blue. The X-ray data from the Chandra X-ray Observatory is shown in pink.

El Gordo is actually two galaxy clusters in the process of merging. One sees separation between the hot X-ray emitting gas (pink) and the dark matter (blue). The hot gas is slowed down by frictional processes during the merger, but the dark matter is not, so we see it as more extended.

Chapter 5

The Great Telescopes

The Great Telescopes are both Earth-bound and in orbit. Here are a few of the great telescopes that are allowing us to explore the universe to greater and greater depth and with increasing sensitivity, in the optical, the infrared, the radio, and in ultraviolet and X-ray wavelengths.

Great Telescopes on Earth

Because of interference from the atmosphere and from terrestrial sources of radio signals, ground-based astronomy is performed mainly from mountaintops (for optical observatories) and desert locations (for radio observatories). Nothing is better than a remote desert mountaintop for an astronomer! Let's look at four of the most important telescopes located on Earth.

By observing in different regions of the electromagnetic spectrum such as optical, infrared, and radio, astronomers can explore different physics and thus better determine the properties of galaxies and other astronomical objects.

Four of the most important ground-based observatories are:

1. Keck Observatory, Hawaii, USA - for optical astronomy

2. Arecibo Observatory, Puerto Rico, USA - for radio astronomy

3. Very Large Array, New Mexico, USA - for radio astronomy

4. ALMA Observatory, Chile - for submillimeter astronomy

Keck Observatory

Image credit: NASA

The W.M. Keck Observatory has two telescopes that are the largest optical telescopes in the world. Each telescope has a massive 10-meter diameter mirror. Like all modern telescopes, each mirror is composed of many segments, in this case 36 hexagonal segments for each mirror.

The Keck Observatory is located on the Big Island of Hawaii, on the summit of Mauna Kea, with superb viewing conditions. It is operated by the California Institute of Technology and the University of California, although researchers from around the world can submit proposals for observing time. The twin telescopes were constructed with a $70 million grant from the Keck Foundation.

Operations at Keck began in 1990. A variety of instruments can be used in conjunction with the telescopes, for example the DEIMOS instrument (The Deep Extragalactic Imaging Multi-Object Spectrograph) can take spectra from 130 galaxies simultaneously!

Arecibo Observatory

Image credits: By H. Schweiker/WIYN and NOAO/AURA/NSF

For many years the largest single dish radio telescope used for astronomy has been the Arecibo telescope located in Puerto Rico and initially completed in 1963. (A new larger Chinese telescope is just coming on line). The 1000 foot (305 meter) diameter dish is built in a natural valley, and is an altazimuth design.

Because of its appealing design, the Arecibo Observatory was featured in the James Bond film <u>Golden Eye</u> and also in <u>Contact</u> and in <u>Species</u>, as well as in an <u>X-Files</u> episode.

A 900 ton superstructure hangs above the primary, suspended from three large concrete towers, and is used to support the feed where the radio waves come to a focus. Only objects within 20 degrees from the azimuth can be observed, by moving the hemispherical feed along the a metal track.

A major upgrade of the surface occurred in 1974 to replace the original wire mesh with 40,000 aluminum panels, configured as a section of a sphere. This allowed for observations at higher frequencies and with better resolution.

The author used this radio telescope as a graduate student for research on quasars and walked around on the catwalk that is suspended in mid-air, providing an excellent view of the Caribbean Sea. Both exhilarating and scary - especially in the middle of the night!

Very Large Array

Image credit: NRAO/AUI and Photographer - Kelly Gatlin; Digital composite-Patricia Smiley

The Very Large Array (VLA) consists of 27 radio telescopes with diameters of 25 meters located at a single site west of Socorro, New Mexico. The telescopes can operate in unison by using a technique known as aperture synthesis.

The signals from the various telescopes can be combined as if they were a single very large telescope. The collective array has 27 times the radio gathering power, similar to a single dish over 5 times larger in diameter (over 125 meters) and 27 times larger in surface area.

The spatial resolution of the array is enhanced considerably by spacing the telescopes at various distances. The telescopes are able to move on rails to change their configuration.

Radio astronomy with the VLA allows the study of a variety of phenomena, including active galactic nuclei and mapping of the hydrogen gas distribution in our Milky Way galaxy.

The VLA is featured in several movies, including <u>Contact</u>.

Atacama Large Millimeter Array

Image credit ALMA (ESO, NAOJ and NRAO), C. Padilla

The Atacama Large Millimeter Array (ALMA) is an array of telescopes for millimeter wave observations located in the Atacama desert in Chile, a very high quality observational location. It is the most expensive ground-based telescope in existence, at around $1.4 billion. ALMA is an international partnership including Europe, the U.S., Canada, Japan, South Korea, Taiwan, and the Republic of Chile.

Around half of the ALMA antennas are shown in this photo; most are 12 meters in diameter. Like the VLA, aperture synthesis is used to combine the signals from the various telescopes. ALMA has been in operation since 2013. The initial array has 66 telescopes, and the telescope separations can be as large as 16 kilometers. ALMA operates at much shorter wavelengths (higher frequencies) than the VLA, over the range from 1/3 millimeter to 10 millimeters wavelength.

Great Telescopes in Space

In space, there is even more freedom to observe light from different regions of the electromagnetic spectrum (radio, infrared, optical, ultraviolet, X-rays and gamma rays), allowing astronomers to explore a range of physics and the properties of galaxies, stars and gas. Extragalactic astronomy and cosmology are major research areas for all of these space-borne telescopes.

In this section we look at five of the most important telescopes orbiting in space (in one case, planned for future launch). Because of interference from the atmosphere and from terrestrial sources of signals, ground-based astronomy is performed mainly from remote locations. In the last section we said nothing is better than a remote desert mountaintop for an astronomer, but even better than that in many cases is a telescope in orbit, far removed from the Earth's atmosphere and most sources of interference.

There have been many astronomy satellites or astronomical experiments hosted in space. Any short list of the most important is arbitrary, but we have chosen five major astronomy satellites and instruments, including three of the four NASA Great Observatories:

1. Hubble Space Telescope, a NASA Great Observatory, launched 1990, optical

2. Spitzer Space Telescope, a NASA Great Observatory, launched 2003, infrared

3. Chandra X-ray Observatory, a NASA Great Observatory, launched 1999, X-ray

4. Planck space observatory, European Space Agency, operation 2009 - 2013, microwave

5. James Webb Space Telescope, NASA, scheduled for 2018 launch, infrared and optical

Hubble Space Telescope

This is the most famous of telescopes, and everyone has seen many of the stunning images of galaxies and other objects from the Hubble telescope (HST). The HST has a 2.4 meter primary mirror and is able to make observations in the near-infrared, visible and ultraviolet portions of the spectrum using five main instruments.

Edwin Hubble was the most famous astronomer of the 20th century. Using the Mt. Wilson Observatory in California, he discovered in 1929 that distant galaxies are all moving away from us, due to the expansion of the universe. This was the first observational evidence for the Big Bang theory, and the parameter that defines the expansion rate and the approximate age of the universe is named for him and known as the Hubble constant.

The Hubble Space Telescope (HST) is named in his honor and has contributed greatly to the extragalactic astronomy that he pioneered.

The Hubble Space Telescope had a difficult early history. It was launched via the Space Shuttle in 1990, but found to have aberration in the optics; this was a problem for imaging faint extended objects such as distant galaxies. Corrective optics were installed in a servicing mission from the Shuttle Endeavour in 1993, and it has operated brilliantly since that time, long past its expected life. Several servicing missions have upgraded the instrumentation. Its successor, the James Webb Space Telescope, is scheduled for a 2018 launch.

Spitzer Space Telescope

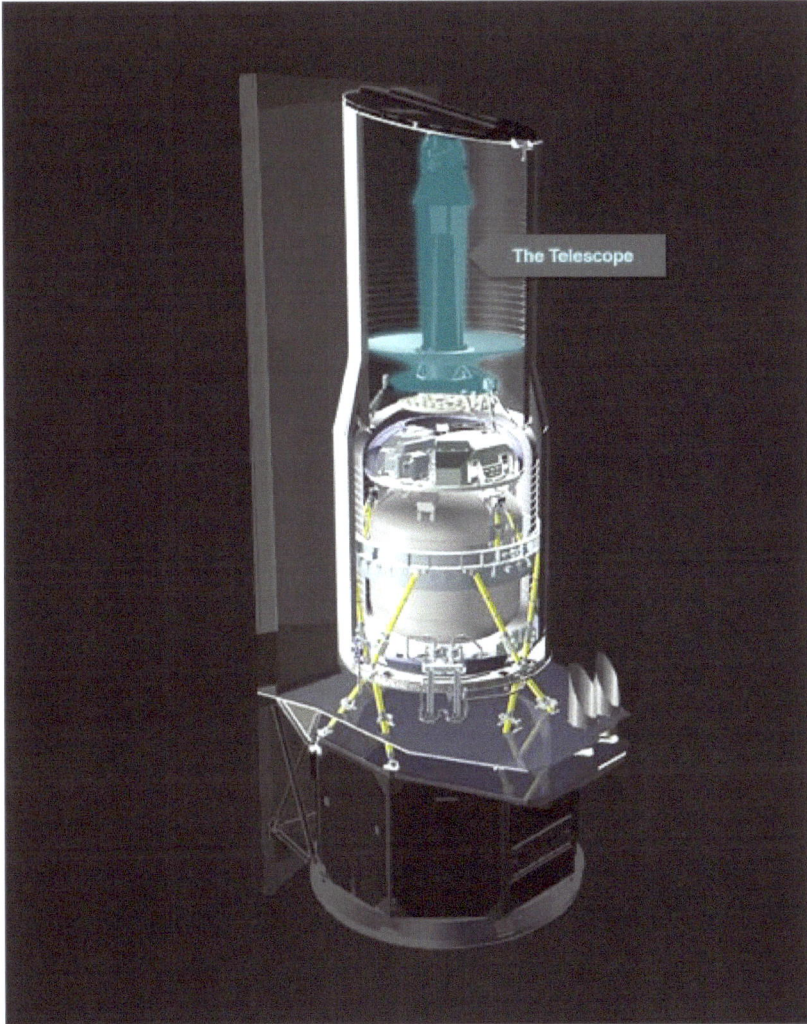

Image credits: Spitzer Telescope: NASA/JPL-Caltech; Lyman Spitzer public domain

The Spitzer Space Telescope is designed to detect sources of infrared radiation. In our galaxy, cool stars and cold, dense molecular clouds, which are sites of star formation, are of great interest for infrared astronomy. Infrared telescope are also useful for observations of very distant galaxies, since their light is shifted from optical wavelengths into the infrared, due to the redshift effect of the expanding universe. For example, galaxies at redshift of 1 have their light shifted to half the frequency or twice the wavelength as compared to nearby galaxies. Galaxies that are undergoing rapid bursts of star formation are of particular interest as infrared sources at high redshifts.

Observations from the Hubble Space Telescope and from the Spitzer Space Telescope can be compared to extract even more information about galaxies.

The Spitzer Space Telescope is named in honor of Lyman Spitzer, one of the pioneers in studying the interstellar medium and star formation. It was launched into an 'earth-trailing' heliocentric orbit in 1993.

The telescope is made of lightweight beryllium and was cooled with liquid helium until 2009, when the supply was exhausted, so most of the instruments are no longer usable. Useful data is still being obtained with the IRAC camera on board.

The author of this book was a doctoral student of one of Spitzer's doctoral students (George Field).

Chandra X-ray Observatory

Image credits: NASA/CXC/NGST; http://chandra.harvard.edu/resources/illustrations/ chandraPortraits.html

X-rays are heavily absorbed by the Earth's atmosphere, so X-ray astronomy requires placing X-ray telescopes in the upper atmosphere with balloons or in space on satellite platforms. X-rays also penetrate glass-based optics, so X-ray mirrors are of the grazing incidence type, made of highly polished metal.

There are many X-ray sources in the cosmos, including neutron stars and black holes (the hot gas nearby black holes, actually) and active galactic nuclei powered by supermassive black holes. Rich clusters of galaxies are filled with very hot gas that radiates in the X-ray portion of the spectrum.

One of the most important X-ray observatories is Chandra, named for the great Indian astrophysicist Subrahmanyan Chandrasekhar. He was primarily based at the University of Chicago and worked on a wide variety of problems ranging from stellar structure and evolution to general relativity; he received the Nobel Prize in Physics in 1983 for his work on stellar evolution.

The white dwarf upper mass limit of 1.4 solar masses was worked out by him and is named in his honor. It turns out to be quite important since white dwarfs that accrete mass above this limit will "go supernova" and supernovae of this type all have more or less the same luminosity. These Type Ia supernovae work as standard candles to help us tie down the absolute distance scale to galaxies.

The Chandra X-ray Observatory was launched as a payload of the Space Shuttle Columbia in 1999 and continues to operate today. At the time of launch it was far and away the most sensitive X-ray telescope, and is credited with many discoveries and advances in X-ray astronomy, including the first X-ray detection of the supermassive black hole at the center of our Milky Way.

Planck Satellite

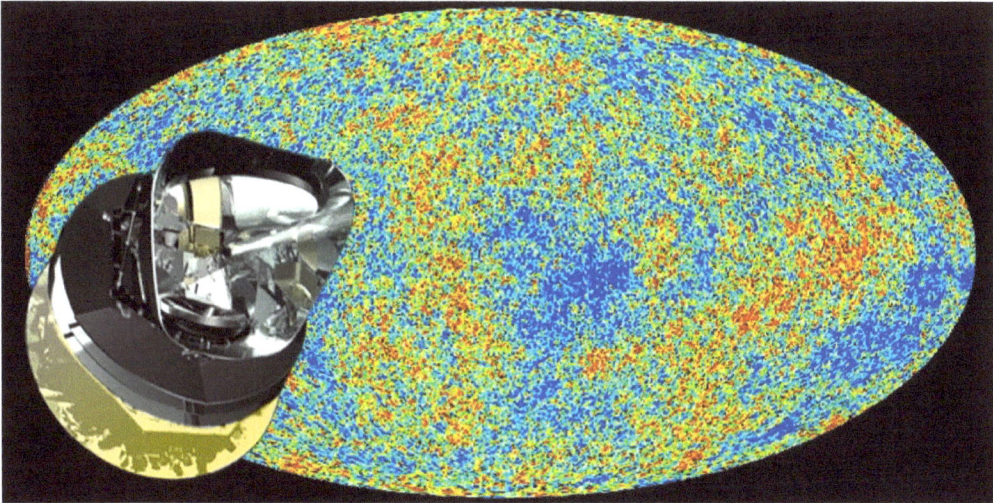

Image credits: Planck satellite and CMB map, European Space Agency; Max Planck photo, public domain

One of the most important astronomical observatories is known as Planck, named for the famous German quantum physicist Max Planck. The Planck satellite, now turned off, was launched by the European Space Agency in 2009 to observe the cosmic microwave background (CMB) radiation; it was placed at the stationary L2 Lagrange point (see Glossary) relative to the Earth's orbit about the Sun. It was designed to detect photons in the microwave and far-infrared portions of the spectrum.

The CMB is the remnant radiation dating from a time when the universe was 380,000 years old. Its discovery in 1965 proved the case for the Big Bang model. The image shows the (very small) temperature fluctuations in the CMB as measured by Planck. These reflect density fluctuations which are thought to be due to the original quantum processes that created the universe. The overdense regions, hundreds of millions of years later, would begin the formation of the first galaxies and stars.

Major data releases in 2013 and 2015 support the canonical Big Bang cosmological model with a universe dominated by dark energy and dark matter. Ordinary matter only accounts for 5% of the mass-energy balance of the universe. Not only is the universe expanding, but since dark energy has twice the strength of dark matter and ordinary matter combined, the expansion of the universe is accelerating. The form of the dark energy matches Einstein's cosmological constant in the equations of general relativity. What Einstein called his "biggest blunder" seems not to have been a blunder at all!

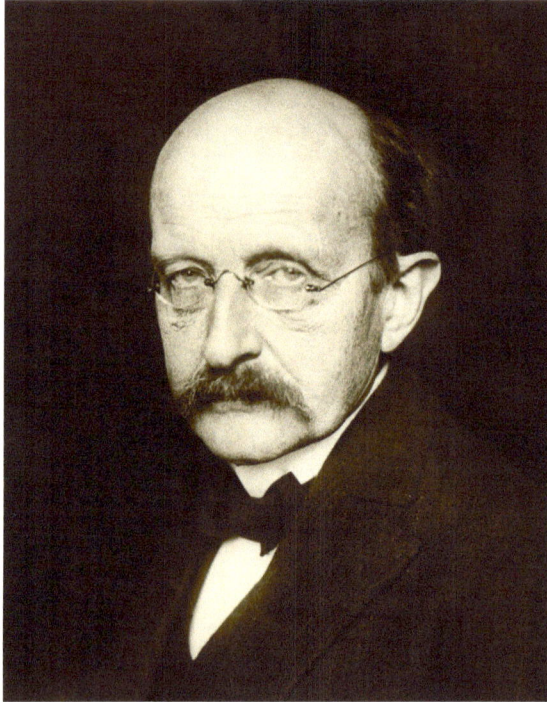

James Webb Space Telescope

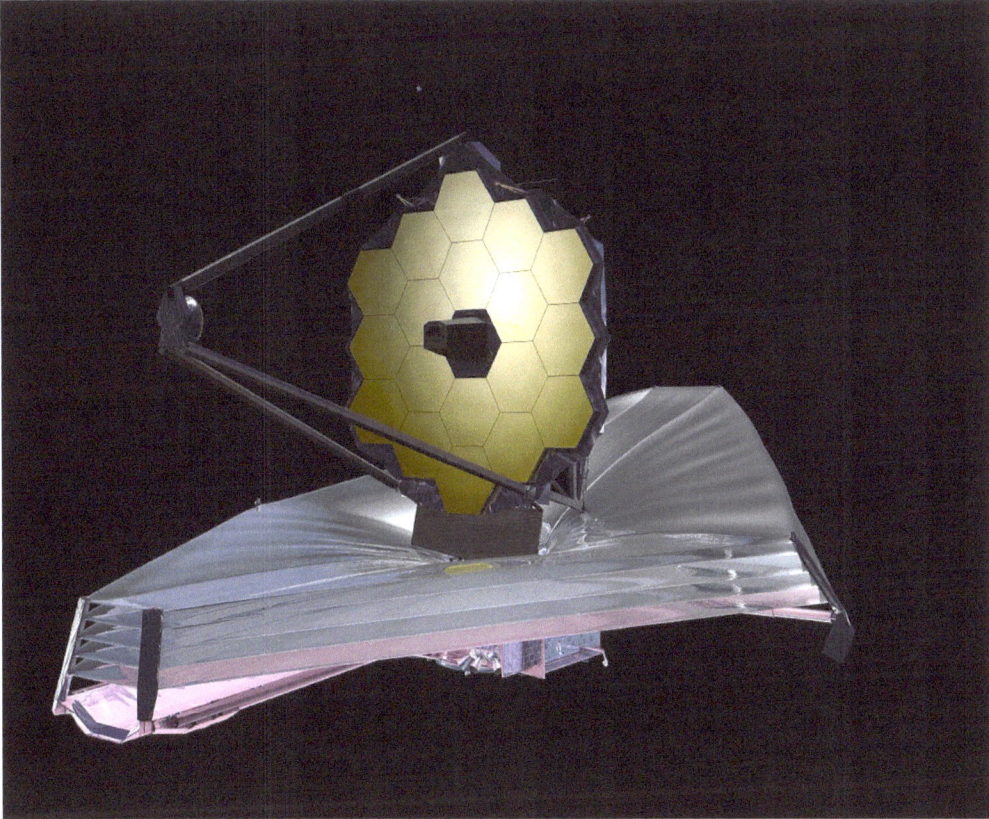

Image credit: NASA

The James Webb Space Telescope is the planned successor to the Hubble Space Telescope, and is currently being assembled and prepared for launch in 2018. It is designed for both optical and infrared observations, so serves as a successor to the Spitzer Space Telescope as well.

The telescope is 5 times larger by area than the Hubble Space Telescope, with a 6.5 meter diameter as compared to Hubble's 2.4 meter size. There is a large sunshade to protect the telescope from sunlight and allow it to be kept cool for best observations in the infrared portion of the spectrum. The telescope will be positioned near the Earth-Sun L2 Lagrange point.

The project is led by NASA with significant contributions from the European Space Agency and Canada.

Because it is designed to observe well into the infrared, it is highly suited for studies of extremely distant galaxies, which have heavily redshifted spectra due to the cosmological expansion. For example, a galaxy with a redshift of 9 (very early universe, only half a billion years old) will have its Hydrogen alpha emission shifted ten times from .656 microns (red) to 6.56 microns (well into the infrared). And the Lyman alpha emission signature of young galaxies at the same redshift moves from the deep ultraviolet to the near infrared region at 1.2 microns.

There is little doubt that the James Webb telescope will produce an incredible amount of science, including in the search for extraterrestrial planets, the understanding of star formation, and our knowledge of galaxy formation and evolution. It should allow us to see the first generation of stars and galaxies being formed. The telescope is named for James E. Webb, who was NASA's second administrator during the 1960s.

Chapter 6

Closing Thoughts

Image credit: View toward the Galactic Center. ESO/Y. Beletsky, Paranal Observatory

We promised you 72 beautiful galaxies, and we have actually shown you many more than that.

* 15 galaxies, either from the Messier catalog or nearby dwarf galaxies

* 21 galaxies from the NGC catalog

* 12 high-z (high redshift) very distant galaxies, looking back to the earliest years of the universe

* 24 galaxies from Groups

* And also, hundreds of galaxies in Clusters, too many to enumerate

The galaxy shown on the first page of this chapter is the most beautiful of all - it is our home - our Milky Way Galaxy. Our galaxy is home to some 200 billion or more stars. Its morphology is that of a thin disk only 2,000 light-years thick, but with a diameter of over 100,000 light-years. And it contains a supermassive black hole of 4 million solar masses in its center. The total mass of the galaxy is over a trillion times the mass of our Sun.

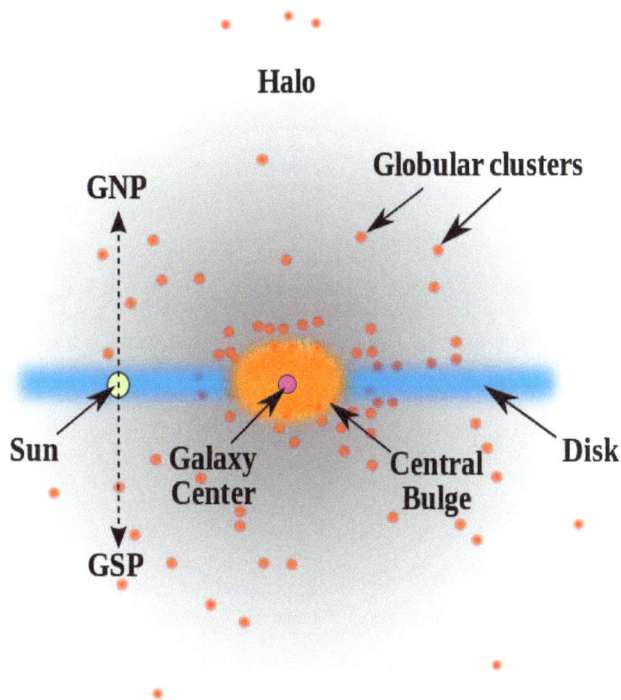

Schematic profile of Milky Way Galaxy; Image credit: R.J. Hall, (CC BY-SA 3.0)

We live in the 'suburbs', some 27,000 light-years from that center, in the Solar System. Our Sun revolves around the Milky Way's center each 240 million years or so. The Sun, which sustains all life on Earth, is around 4.6 billion years old, but the oldest known star in our galaxy is over 13 billion years of age, nearly the age of our universe, which is 13.8 billion years.

The total mass of the galaxy is dominated by the mysterious dark matter that extends in a halo well above the disk. Yet the dark matter density in the Sun's neighborhood is very low; it is on average the equivalent of less than one proton per cubic centimeter. Dark matter is not protons or neutrons or electrons, it is some yet-to-be-discovered exotic particle or possibly black holes formed in the very early days of the universe. Yet the gravitational effects of dark matter are beyond dispute.

We've taken you on a trip from relatively near to very far away in the universe, with 15 images of "M" galaxies, 18 images of "N" galaxies (plus 3 found in pairs), 11 "z" galaxies (plus a companion), and 24 galaxies found within 6 different groups. That's 72 galaxies - but beyond that you have seen hundreds of galaxies that are found within 7 rich clusters. You have seen that galaxies come in many shapes and sizes, with major categories being spirals and ellipticals. There are also many irregular galaxies and galaxies undergoing mergers or being cannibalized by other galaxies. Galaxies vary tremendously in their masses, in their luminosities, in their shapes, in their star formation rates, and in the amount of gas and dust they contain. And some are near, in cosmic terms, and some are billions of light-years away, at the edge of the universe. Many are found in groups, including our own Milky Way. Others are found within very large clusters.

I hope this book has provided you with some appreciation of the variety of galaxy types, as well as their beauty, and some feeling for what astronomers are seeking to understand as we explore and discover more and more galaxies with the Great Telescopes of our day. Exactly how galaxies form and evolve is a very active area of current research. Advances are being made rapidly, but there is still a great deal that is not yet understood.

In the decades to come, even more powerful radio, infrared, optical, ultraviolet, X-ray and gamma-ray telescopes, as well as gravitational wave detectors, will help us further unlock the secrets of the galaxies. And this will be done in concert with more sophisticated theoretical models of galaxy formation and evolution, supported by detailed simulations with ever more powerful supercomputers.

Glossary

Andromeda Galaxy - M31, or the Andromeda Galaxy is the best-known external galaxy. It is twice or more the size and mass of the Milky Way, and is moving toward our galaxy. It is expected that the two will collide in about 4 billion years.

Big Bang - The Big Bang is the best established model for the universe's formation and history over the past 13.8 billion years. The universe has been expanding continually since its creation, as evidenced by the recession of the galaxies from one another on large scales, and by the cosmic microwave background radiation left over from the Big Bang.

Black Hole - An object with a very high density such that neither matter nor light can escape its clutches, in accordance with general relativity. Typical stellar mass black holes are of 3 to 10 solar masses. A 3 solar mass black hole has an event horizon of only 10 kilometers. There are also many supermassive black holes of much greater mass.

Cluster of Galaxies (galaxy cluster) - A cluster of galaxies is a large group with hundreds or even a thousand galaxies or more, bound together by mutual gravitational attraction. Most of the mass in clusters is in the form of the mysterious dark matter. They are usually strong emitters of X-rays due to hot gas at tens of millions or even a hundred million degrees found between the galaxies in the cluster.

Constellation - A group of stars within our own galaxy the Milky Way. We see these together on the sky, but the stars within the constellation will be at various distances away from us. When we say a galaxy is in a certain constellation, we really mean it is in that direction of the sky, but it is of course behind the constellation in reality - since all the constellations are within our galaxy.

Cosmic Microwave Background Radiation - Isotropic blackbody relic radiation from the Big Bang emitted at time 380,000 years after the origin. It is presently at a temperature of 2.73 degrees (Kelvins) above absolute zero, and is seen in the millimeter region of the electromagnetic spectrum.

Dark Energy - Non-matter contribution to the overall energy-mass balance of the universe. It may be due to quantum mechanical vacuum energy or due to a time varying field "quintessence". Observations indicate it takes the simple form of an unvarying cosmological constant, consistent with general relativity. Einstein discovered it theoretically, called it his "biggest blunder", but it was not a blunder. Dark energy currently accounts for 2/3 of the energy-mass balance of the universe. It is also causing the universe to accelerate in its expansion.

Dark Matter - Dark matter has been discovered in galaxies, groups of galaxies, and clusters of galaxies due to its gravitational effects. In the universe as a whole, measurements indicate that there is five times as much dark matter as ordinary matter! It does not radiate light at any frequency. Attempts to directly detect it have so far come up short; it is believed to be composed of an exotic new fundamental particle that does not interact with light, or possibly of black holes formed in the very early universe (primordial black holes).

Edwin Hubble - Edwin Hubble was a great astronomer for whom the Hubble Space Telescope has been named. He discovered the general pattern of the recession of galaxies known as Hubble's law; this states that their velocity of recession away from our galaxy is proportional to their distance. This was the proof of an expanding, isotropic universe and support for the Big Bang theory.

False Color - Using colors that are visible to the human eye to represent data points taken at frequencies that are invisible, such as infrared and X-rays.

Galaxy - A galaxy is a gravitationally bound collection of dark matter, stars, gas and dust. The stars, gas and dust are detectable through the light of all frequencies that they emit (and absorb). The dark matter is detectable through its gravitational influence on stellar orbits. The main types of galaxy are spiral, elliptical and irregular, and spiral galaxies may or may not have a bar. Galaxies also contain planets, comets and asteroids; these are generally found associated with the stars to which they belong.

Globular Clusters - Globular clusters are dense star clusters with very old stars, found in our galaxy and other galaxies. They were formed during the early part of the universe, so tend to have old stars with low metal content.

Gravitational Lensing - A massive foreground object will distort the light paths of background objects, as predicted by general relativity. Often a cluster of galaxies serves as the gravitational lens, and both brightens and distorts the images of galaxies behind the cluster. The distorted galaxies are easier to detect due to the brightening effect, which can be large.

Great Debate - The Great Debate between two prominent astronomers, Harlow Shapley and Heber Curtis was held in 1920. Curtis maintained that some of the smudgy spiral nebulae seen on photographic plates were galaxies (or 'island universes') external to our own, while Shapley said they were clouds of gas and stars within our own galaxy. Curtis was proven right, so external galaxies have been known as such for about 100 years. Immanuel Kant (who gave the designation 'island universes') had held the same opinion in the 18th century, but it was not until 1920 that the observations became definitive. Once these spiral nebulae were determined to be external galaxies, the physical scale of the universe was understood to be tremendously larger.

Group of Galaxies - A group of galaxies is a group found together on the sky. A true group is a bound 3-dimensional set of galaxies, and with a few to several tens of members. Our Local Group includes Andromeda and the Milky Way and another 50 or so galaxies.

Hubble Flow - The Hubble flow is the rate at which galaxies recede from one another due to the expansion of the universe. The so-called Hubble constant relates the speed of recession as being proportional to the distance between the galaxies. Distant galaxies that are farther away from us recede at faster rates. This recession results in a redshift of the light from the galaxies.

Inflation - A period at the very beginning of the Big Bang during which the universe expanded from a tiny sub-microscopic volume up to the macroscopic scale, undergoing many doublings in size. This was powered by release of energy from a decaying 'inflaton' field. The released energy ended up as radiation and also the dark matter and ordinary matter which fills the universe.

Infrared Radiation - Infrared radiation, or light, is redder than red. It has longer wavelengths than visible light. Astronomers talk about the near infrared and the far infrared, extending from about .75 micron to 300 microns in wavelength. (A micron is one ten-thousandth of a centimeter.)

Lagrange Point - A point of equilibrium in the Earth's orbit where a spacecraft may be "parked". The L1 and L2 points are along the Earth-Sun axis, with the L1 point closer to the Sun and the L2 point farther away from the Sun.

Light Year - A light year (or light-year) is the distance traveled by light in one year. Light travels at 300,000 kilometers per second, and 9.45 trillion kilometers per year.

Magnitude - Astronomers use the magnitude scale to denote brightness. Larger numbers indicate fainter objects, smaller numbers indicate brighter objects. Both apparent magnitude and absolute magnitude can be measured. Apparent magnitude is the brightness as we see the object, whereas absolute magnitude is the actual luminosity of the object, correcting for its distance. The magnitude scale is logarithmic, with five steps or magnitudes indicating a factor of 100 in relative brightness.

Messier Catalog - There are 110 Messier objects in the catalog. It was first published in 1771 by French astronomer Charles Messier (his surname is pronounced meh-z-ay), and was a catalog of nebulae and star clusters. It was not known at the time that some of the nebulae were extragalactic objects - galaxies in their own right. It took another 150 years to fully establish that many of the Messier objects are galaxies.

Milky Way - Our galaxy is known as the Milky Way; this also refers to the band of unresolved stars that we can see at night. We are looking through the plane of our galaxy, which is a barred spiral, when we see those stars. There are thought to be over 200 billion stars in our Milky Way.

Neutron Star - A stellar remnant composed primarily of neutrons, formed after a massive star goes through a supernova phase. If the remnant is above 2 to 3 solar masses, the neutron pressure is insufficient and the remnant will collapse to a black hole.

NGC catalog - NGC, or the New General Catalog, was compiled in 1888 by Jean Louis Emil Dryer. It contains 7,840 objects. Many of the objects are galaxies, but many are nebulae or objects within our own Milky Way.

Quasar - A quasar, or quasi-stellar radio source, is a galaxy with an active galactic nucleus. The nucleus will typically outshine the rest of the galaxy and can be up to 100 times as bright or more as our entire Milky Way Galaxy. They are generally discovered in the radio or optical bands. Often these active galaxies will emit radiation in the ultraviolet, X-ray and gamma ray bands as well. The energy comes from an accretion disk surrounding a supermassive black hole in the center of the galaxy.

Radio Waves - Radio waves are actually another form of light, with longer wavelengths than infrared radiation. Everyone is familiar with frequencies in the Kilohertz and Megahertz (MHz) bands from radio stations. Much radio astronomy also happens in the Gigahertz bands, at higher frequencies. The famous 21 centimeter line of hydrogen emission and absorption is at 1420 Mhz (1.42 GHz). This book was written with a 3.5 GHz processor. We detect radio waves with radio telescopes (often parabolic reflectors with metallic surfaces) and arrays, rather than optical telescopes that use glass elements.

Redshift - The shift of light emitted from distant galaxies toward the red (and infrared) ends of the spectrum, due to the galaxies' movement away from us as the universe expands (see Hubble flow). Astronomers use the symbol 'z' to represent the value of redshift. A redshift of z = 0 implies no shift, and redshifts for galaxies are now being observed up to z = 11. A few galaxies nearby have blueshifts due to their peculiar velocities and because they belong to our Local Group, but almost all galaxies in the universe display the redshift effect. If the redshift is z = 1, the wavelength of light is doubled from what it would be in a laboratory on Earth.

Spiral Arms - Spiral galaxies are so designated due to their two or more arms which extend out from the center and curve gracefully in a spiral fashion. The spiral arms are density waves - areas of enhanced density of stars and gas. The stars and gas can pass through the spiral arms, but they spend relatively more time in the spiral arms, just as you spend relatively more time on the freeway moving through areas of high traffic that slow your progress.

Supermassive black holes - Most black holes are of stellar mass, a few times the mass of the Sun. Supermassive black holes are found at the centers of galaxies, have masses from a million up to a billion or more solar masses. They power highly energetic active galactic nuclei, such as quasars.

Supernova - A massive star at the end of its life. A supernova explosion has very rapid thermonuclear fusion and can temporarily be as bright as an entire galaxy. The star's outer layers will be thrown out into the interstellar medium, but the core will remain in most cases as either a neutron star or black hole depending on the progenitor star's mass.

Tuning fork diagram - Edwin Hubble's classification scheme for most galaxies. Both spiral galaxies (S) and elliptical galaxies (E) are represented. The spirals may be barred (SB) or not (S), thus the "tuning fork" shape. Numbers 0-7 indicate the flattening in the case of ellipticals.

Ultraluminous infrared galaxy (ULIRG) - An infrared galaxy is a galaxy that is very distant, so that its visible light is shifted into the infrared. Using infrared telescopes on the ground and in space is an excellent way to detect high-redshift galaxies, that is, galaxies as they were in the early universe. An ultraluminous infrared galaxy is especially bright because of high star formation rates, perhaps 100 times or more greater than our Milky Way's current rate.

Ultraviolet light - Ultraviolet light has higher frequency and thus shorter wavelength than visible light. It sits in between visible light and X-rays in the electromagnetic spectrum.

White dwarf - The Sun will end up as a white dwarf. These are stellar remnants formed at the end of life after a star has completed all of its possible thermonuclear fusion and exhausted the fuel in its core. The white dwarf is held up by the pressure of its electrons and can have a maximum mass of 1.4 solar masses, above which it will collapse to a neutron star.

X-rays - Most of us are familiar with X-rays from the dentist's office. X-rays can range in energy from under 1 KeV (KiloVolt) to over 100 KeV. Because of the high energies of X-ray photons, X-ray telescopes use grazing incidence techniques, or an aperture grille in front of a detector such as a CCD. X-ray telescopes for astronomy must be airborne in balloons or placed on satellites to be above the Earth's atmosphere, which blocks most X-rays.

References and Suggested Reading

Dark Matter, Dark Energy, Dark Gravity (2nd edition), Stephen Perrenod, 2013 (available on Amazon in paperback and Kindle formats). This book provides an introduction to modern cosmology, including the Big Bang, inflation, dark matter, dark energy and the mystery of gravity.

Meet the Frontier Fields http://frontierfields.org/meet-the-frontier-fields/ "Frontier Fields draws on the power of massive clusters of galaxies to unleash the full potential of the Hubble Space Telescope."

Videos

https://www.youtube.com/watch?v=zN3CUWqPuhE - video of all of the Messier objects

https://en.wikipedia.org/wiki/Andromeda%E2%80%93Milky_Way_collision - video of a simulation of Andromeda Galaxy and the Milky Way colliding, several billion years in the future

About the Author

Image credit: Napapat Kraisophon

Stephen Perrenod holds Ph.D. and Master's degrees in Astronomy from Harvard University and a Bachelor's in Physics from M.I.T. His primary research focus was on the cosmological evolution of X-ray emitting clusters of galaxies. He later worked in the high performance computing (HPC) field and has been a frequent public speaker on HPC and Cloud computing topics and a consultant in Big Data, IoT, and Deep Learning. He is the author of Dark Matter, Dark Energy and Dark Gravity, a cosmology book about the mysterious nature of dark matter and dark energy and the elusiveness of our understanding of gravity. His blog is at http://darkmatterdarkenergy.com and Dark Matter, Dark Energy, Dark Gravity is available on Amazon at:

https://www.amazon.com/Dark-Matter-Energy-Gravity-Intelligent-ebook/dp/B00F1RQZPQ/

www.ingramcontent.com/pod-product-compliance
Lightning Source LLC
Chambersburg PA
CBHW041444210326
41599CB00004B/125